THE
INTERESTING GOLDEN RATIO

THE
INTERESTING GOLDEN RATIO

A Simple Mathematical Approach

Vincent Siu

PARTRIDGE

Print information available on the last page.

To order additional copies of this book, contact
Toll Free 800 101 2657 (Singapore)
Toll Free 1 800 81 7340 (Malaysia)
orders.singapore@partridgepublishing.com

www.partridgepublishing.com/singapore

Contents

To my parents

Chapter 1

Introduction

1.1   Fascinating Ratio

The golden ratio, or golden section, has been a fascinating, or even regarded as mysterious proportion in mathematics, architecture, art, archaeology and botany for centuries. The earliest written study of the proportion can be found in *The Elements*, a mathematical and geometric treatise written by the ancient Greek mathematician Euclid, 2,300 years ago. Euclid named it as the Mean and Extreme Proportion and proposed construction methods and proofs.

After about 1,800 years, Luca Pacioli (1445 – 1517), an Italian Mathematician, followed the geometry of Euclid in his book *Divine Proportione* (On Divine Proportion), published in 1509. He studied and explained the applications of the golden ratio which he called the divine proportion to the regular polyhedrons, visual art and architecture. Illustrations in the book were drawn by Leonardo da Vinci. It is Leonardo who first named the ratio as the *section aurea* (Latin for golden section). Evidences have been discovered to support that the ancient Egyptians had incorporated this proportion in producing earthenware and building pyramids in the period 2,000 – 3,000 BC. So are constructions and temples left by ancient Greeks, notably the Parthenon, temple to the goddess Athena, in the Acropolis in Athens, Greece. Arts and crafts of the Renaissance period are said to have applied the ratio to enable paintings and architecture appear more aesthetic.

The rectangle constructed in the golden ratio, known as the golden rectangle, is said to be most aesthetically pleasing, and being regarded as the shape of beauty. It has been used in the outlines and facades of buildings, and designs and outfits of commodities from the Renaissance up to twentieth century. The world renowned architect Le Corbusier applied the golden rectangles as the shape of windows and other dimensions in buildings he designed. He introduced two series of architectural proportions based on the golden ratio. Even the United Nations Headquarters in New York is quoted by many as composing of the golden section in the tall and low wings. The shape of a credit card is regarded as a golden rectangle (measurement shows it is near to but not exactly). The growth of plants is noted to be in patterns according to the golden ratio or Fibonacci Numbers, the two being closely related mathematically. Renaissance astronomer Johannes Kepler (1571-1630) have proved that the ratio of consecutive Fibonacci Numbers approaches the golden ratio. We shall go into details between the Fibonacci Numbers and the golden ratio in Chapter 6.

Figure 1.1 <u>United Nations Headquarters</u>
(Source: Pixabay)

Many writers have argued that famous composers like Mozart and Beethoven applied the golden ratio in their music, making the rhythms more pleasing. In the same token, it has been proposed that Virgil and other ancient Roman poets deliberately used the Fibonacci numbers to frame their poems. While analysis of the relationship between the golden ratio and music or poetry is beyond the scope of this small book, interested readers could search for their answers in the reference.

Mathematically, the symbol of the golden ratio is taken as ϕ (φ), a Greek alphabet pronounced as 'phi'. It is said to have been attributed to Phidias (480 – 430 BC), a Greek sculptor who designed the statues of the goddess Athena in the Parthenon and the Parthenon itself. ϕ is numerically approximate to 1.618 for ratio >1, or 0.618 for ratio <1. It will be shown in Chapter 3 how this is arrived, and for the time being, just take it for granted.

Note that $\dfrac{1}{1.618} = 0.618$

1.2 Ancient Architecture and the Golden Ratio

The golden ratio or the golden rectangle, that is, a rectangle formed with the length and width in the ratio of ϕ : 1, is said to have incorporated into many of the ancient Greek buildings and temples. The most quoted one is the Parthenon, in the Acropolis in Athens. The Parthenon is the temple for the goddess Athena, built around 430 BC. Figure 1.2 shows the front view of the temple. The ratio of the width W and height H is in the golden ratio. Also, the height of the columns from the base h_2 to the remaining portion h_1 is ϕ : 1.

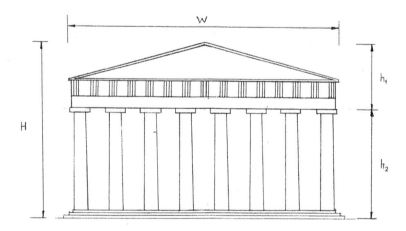

Figure 1.2 <u>Front View of the Parthenon</u>

Figure 1.3 <u>The Parthenon, Athens, Greece</u>
(Source: Pixabay)

The Great Pyramid of King Cheops of ancient Egypt at Giza with an estimated volume of 2,600,000 m³, has a square base. It was discovered that the ratio of one side of the base to the height of the pyramid is about 1.618. In the nineteenth century, Sir John Herschel (1792 – 1871) had translated one of the saying of the ancient Greek historian Herodotus about Egyptian pyramids that "each face of a [square base] pyramid equals to the square of its height". Later we will see that this lead to the conclusion mathematically that the ratio of the altitude of the face to half side of the base is a golden ratio. Another significant discovery is the tomb of Pharaoh Ramses IV, built around 1200 BC, also exhibits the golden ratio. The burial chamber of the Pharaoh consists of 3 rectangular compartments. The innermost one is a double square (rectangle of 1 : 2), the middle one is a golden rectangle (rectangle of the golden ratio), and the outermost one is a double golden rectangle.

Figure 1.4 <u>Egyptian Pyramid</u>
(Source: Pixabay)

Other ancient or millennium old architecture and buildings have been claimed to have use the golden ratio in their construction. For example, the Islamic Great Mosque of Kairouan built in 670 AD, is said to have applied the ratio in the floor plans, court areas and minaret elevations. In Central America, the Maya civilization flourished between 250 to 900 AD, and continued to exist until the invasion of the Spanish in the 16th century. They had constructed quite a number of pyramids of their style. In Mayan language, the pyramid is "witz", meaning mountain. They believed that mountains housed the souls of their ancestors. In the low land where there is no mountain, they built artificial ones. One of the best known Maya pyramids is the EL Castillo (Spanish for castle) situated in Chichen Itza. It has a height of 30m. Also named as the Temple of Kukulcan, it consists of series of square terraces with stairs up each of the four sides. A serpent was carved running down the north balustrade. During the equinoxes, beams of the afternoon sun fall on the northwest corner of the pyramid, casting rows of triangular shadows against the north balustrade, inspired an illusion of a feathered serpent crawling down. John Pile (contemporary scholar) claimed that the interior layout of the El Castillo has golden proportions. The interior walls were said to be placed such that the outer spaces are related to the central chamber by the golden ratio. Some of the designs of other Maya pyramids have been proposed to follow the golden ratio.

Figure 1.5 <u>Maya El Castillo Pyramid</u>
(Source: Pixabay)

The Andean architecture in Southern America can be dated back to about 4000 BC. Andean cultures cover today parts of Ecuador, Colombia, Chile, Argentina, Peru and Bolivia. The long span of land, from present day Quito to Santiago, was then ruled by the Inca Empire by the early 16th century. Stone carvings, temples, mounds for public structures and residences, sunken courtyards and large plazas have been found or unearthed. The Great Pyramid (artificial mountain) at Caral is still 30 m tall. It was constructed around 2800 BC. The pyramid at Moche, Huaca del Sol or Pyramid of the Sun, was originally 50 m in height. It is now 41 m in its ruined form and dated between 1 to 500 AD. Most likely, some golden proportions may be "discovered" in these ancient architecture. The ancient Andeans are also well-known for their ceramic crafts, cotton and woven (from hair of camelids) weaving, with very creative motifs (which may embed some golden ratios).

Figure 1.6 <u>Inca Pyramid</u>
(Source: Pixabay)

1.3   Renaissance Art

Paintings from famous renaissance artists are said to have used the golden ratio in their works. The most controversial one is *Mona Lisa* by Leonardo da Vinci. It measures 77x53 cm (1 : 1.45), not a golden ratio. But there have been numerous analyses that the composition, layout and placement of the main subject, etc. follow the golden proportion. The portrait is commonly regarded as that of Lisa La Gioconda, the young wife of a wealthy Florentine silk merchant Francesco La Gioconda. Lisa's "mysterious" smile is quite usual in portraiture of that period, it appears in Leonardo's other portraits like that of St. Anne and of the Virgin. A contented and modest smile from a woman was considered as a reflection of her beauty and virtue. Afterall, Francesco was a caring husband and Lisa had a good marriage. Already in those days it was good reason for smiling out to the viewers.

Figure 1.7 *Mona Lisa*
(Source: Pixabay)

Livio Mario (2002) remarked that *Mona Lisa* "has been the subject of so many …. contradicting …speculations that it [is] virtually impossible to reach any unambiguous conclusions" with respect to the golden ratio. However, other paintings of Leonardo have also taken as exhibiting the ratio. Both versions of *The Virgin of the Rocks* show a frame ratio of 1:1.63 (199x122 cm and 195.5x120 cm respectively). The main subjects are said to house in a golden triangle.

In 1489, Leonardo took systematic body measurements of two young men. He compared his studies to the theory proposed by Vitruvius, an architect and engineer of the Roman Empire. According to Vitruvius, if a man with outstretched arms and legs would fit into a circle and square (with same centre), then the navel would coincide with the centre. Leonardo corrected the treatise by shifting the square down (Fig 1.8). The drawing is known as the *Vitruvian Man*. Many writers use this figure to show that the human body proportions are closely related to the golden ratio.

Michelangelo, another eminent renaissance artist, painted *the Creation of Adam* on the ceiling of the Sistine Chapel, inside the official residence of the Pope, Vatican City. In the picture, the finger of God touches the finger of Adam exactly at the golden ratio (Fig 1.9). Michelangelo's *Holy Family* is said to have position the family in a pentagram or golden star. His sculptures are also composed of golden ratios. In *The Birth of Venus* by Bottocelli, the canvas is in a golden rectangle. The navel of Venus is at the golden ratio of her height, as well as the height of the painting (Fig 1.10). The last examples given here are Raphael's *Crucifixion* and *Betrothal of the Virgin*. The main characters are said to be enclosed in golden rectangles.

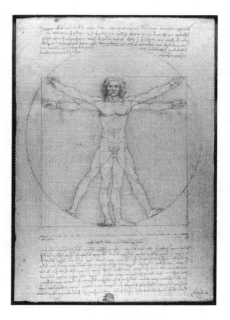

Figure 1.8 *Vitruvian Man*
(Source: Pixabay)

Figure 1.9 *The Creation of Adam* (part of)
(Source: Pixabay)

Figure 1.10 *The Birth of Venus*
(Source: Pixabay)

1.4   Architecture and Art from the Middle Ages to Present

This period excludes that covered by Section 1.3, the Renaissance. The Medieval builders of churches and cathedrals adopted or based on the Greek's designs of temples and halls. Hence inside and out, they were constructed with the golden section. Notable examples are Chartres Cathedral (built 1200 – 1220) and Notre-Dame in Paris (built 1163 – 1250), (Fig 1.11). Chartres is located 70 km southwest of Paris. In the 18th century Rome, there was a revival of classic, mainly Greek architecture. This had influence the western world from late 18th to middle 19th century. Neoclassicism had been introduced in America by Thomas Jefferson with his State Capitol building (1785 – 89) in Richmond, Virginia. The building was an imitation of the Maison Carree, a Roman temple of the late 1st century BC, in Nimes, France. Greek and Roman architecture has long been regarded as incorporating the golden ratio.

Figure 1.11 <u>Notre-Dame de Paris</u>
(Source: Pixabay)

Unlike former masters, the applications of golden section by modern practitioners are usually more certain. Le Corbusier (1887 – 1965) developed a series of standardized human proportions known as the Le Corbusier Modulor. This composes of two series - Blue and Red of architectural proportions based on the golden ratio. The CN Tower in Toronto has a total height of 553.33 m, and divided by the height of the observation deck at 342 m is 1.618 (Fig 1.12). Swiss architect Mario Botta uses the golden ratio in his design of private houses. English artist Joseph Turner (1775 – 1851) made use of colour and light in the golden division to indicate the focus of his painting (Fig 1.13). *La Parade* by the French neo-impressionist Seurat (1859 – 1891) composed of many examples of golden proportions. The canvas of Salvador Dali's *The Sacrament of the Last Supper* (1955) is a golden rectangle. In the picture, a large dodecahedron with edges in golden ratio is suspended above and behind Jesus. Post-modern artist Tony Ashton (born 1948) uses the golden section in his acrylic painting, composing the lines and diagonals of quadrilaterals in that ratio.

Figure 1.12 <u>Toronto CN Tower</u>
(Source: Pixabay)

Figure 1.13 *Modern Rome – Campo Vaccino* by Joseph Turner (note the position of the vortex)
(Source: Pixabay)

1.5   Golden Ratio in Everyday Life

Economists have applied the golden section or Fibonacci numbers to today's financial analysis. One form is the Fibonacci retracement (Elliott Wave Principle), which is a mathematical analysis to determine support and resistance levels in stock or exchange markets based on the Fibonacci numbers. If you read an advertisement on watches or clocks, you would probably notice that the hands are usually set at about 9 past 10 or 10 to 2. Well, the hands are said to form a golden rectangle in this way.

Many company or manufacturer trademarks or logos are designed according to the golden proportion. Some films made in the 20th century are said to have their main themes or climax start at 0.382 (1 − 0.618) of the whole duration. The golden ratio is also used to trim or frame photographs for display. In figure 1.14, the line of the mountain tops is set at 0.62 of the whole height from the upper edge. Is the view more pleasing? Look at the thing that you spend almost one-third of your life on, the mattress on your bed. Measure it and you will probably find that it is close to the golden rectangle. In fact, the "golden" mattress could be 115x186 cm or 122x197.5 cm.

Figure 1.14 <u>Photograph demonstrating the use of the Golden Ratio</u> (by the author)

Chapter 2

Mean and Extreme Proportion

2.1   Fundamentals of Geometry

Before going further, let us recall some basic (Euclidean) geometry.

2.1.1   A straight line is denoted by its 2 end points, say A and B. Also, AB represents the length of the line.

2.1.2   The proportion between 2 straight lines, say AB and CD, is the ratio between their lengths, i.e. AB : CD.

2.1.3   Angle subtended by a straight line is 180°.

2.1.4   The angle between 2 non-parallel straight lines AB and CD is denoted by $\angle AEC$, E being the intercepting point of AB and CD. Note that $\angle AEC$ equals to $\angle BED$ (opposite angles).

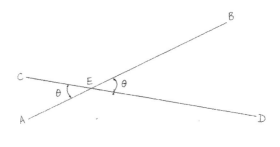

$$\angle AEC = \angle BED = \theta$$

Figure 2.1 <u>Opposite Angles</u>

2.1.5   Corresponding angles and alternative angles.

Refer to figure 2.2, lines AB and CD are parallel. PQ is an inclined line crossing AB and CD. It can be seen that

∠QEB = ∠EFD,   this is known as <u>corresponding angles</u>.

Since   ∠CFP = ∠EFD   (opposite angles)

hence   ∠CFP = ∠QEB   this is known as <u>alternative angles</u>.

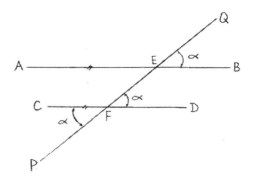

Figure 2.2 <u>Corresponding and Alternative Angles</u>

2.1.6   Two lines are perpendicular to each other if they are at right angle, i.e. 90° apart.

2.1.7   Sum of angles in a triangle is 180°.

Consider an arbitrary triangle ABC (figure 2.3), let the angles at A, B, C be α, β, γ respectively. At point B, draw a straight line EBF parallel to AC.

By alternative angles, $\angle ABE = \alpha$

Also by alternative angles, $\angle EBC = \angle BCD = \alpha + \beta$

At point C, $\alpha + \beta + \gamma$ is angle of a straight line which is 180°

Therefore sum of angles in a triangle equals to 180°.

<u>Remark:</u> $\angle BCD$ is the exterior angle at C, which equals to sum of angles of A and B.

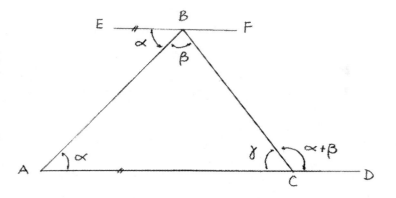

Figure 2.3 <u>Sum of Angles in a Triangle</u>

2.1.8   Angle inscribed in a semi-circle is a right angle (Thales' Theorem)

For triangle ABC, AC lies on the diameter of a semi-circle and B touched its circumference.

Let O be the centre of the semi-circle.

Hence OA = OB = OC, being the radius

Triangle OAB is isosceles (2 sides equal),

$$\angle OAB = \angle OBA = \alpha$$

Similarly, triangle OBC is isosceles, $\angle OBC = \angle OCB = \beta$

$$\angle BOC = \alpha+\alpha = 2\alpha \text{ (exterior angle, refer to Sect 2.1.7)}$$

Similarly, $\angle BOA = \beta+\beta = 2\beta$

$$\angle BOC + \angle BOA = 2\alpha + 2\beta$$

$$= 180° \quad \text{as AOC is a straight line}$$

$$2(\alpha+\beta) = 180°$$

$$\alpha+\beta = 90°$$

As B is an arbitrary point on the circumference of the semi-circle, therefore angle B is always a right angle.

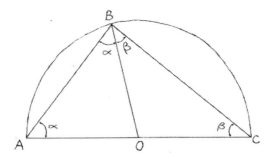

Figure 2.4 <u>Angle inscribed in a Semi-circle</u>

2.1.9   Construct a perpendicular to a line AB

Let O be a point on the line AB. Use a pair of compasses and take O as centre and radius smaller than OA and OB, obtain 2 points C and D on both sides of O. Take a new radius greater than OC (OD). Apply C and D as centres respectively, obtain the intersection of the 2 arcs at E. Draw a straight line from O to E. Then OE is a perpendicular line to AB. The symbol of perpendicularity is ⊥ or ∟.

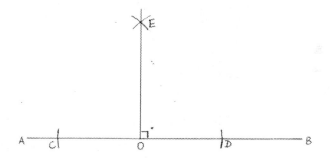

Figure 2.5 <u>Construct a Perpendicular Line</u>

2.1.10   Bisect a line AB (cut into equal halves)

With radius greater than half of AB, take A and B as centres in turn, construct arcs on both sides of AB. Obtain the intersecting points C and D. Join C and D to cut AB at E. Then AE = EB.

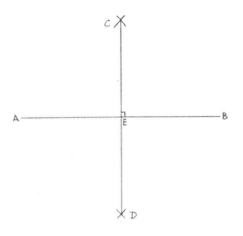

Figure 2.6 <u>Bisect a line AB</u>

2.2   Propositions from *The Elements*

The great Greek mathematician Euclid wrote *The Elements* around 300 BC. It consists of 13 Books and each book contains about 20 – 30 propositions. Geometries relating to the Extreme and Mean Ratio have been mentioned notably in Book VI and Book XIII. We will go through a few of these propositions that are of interest.

2.2.1   <u>Book IV, Proposition 13</u>

*To two given straight lines, find a **mean** proportion.*

Take 2 unequal straight lines AB and CD. The ratio of these 2 lines is simply AB : CD. Let EF be another straight line of length between AB and CD. EF will constitute to a **Mean Proportion** if AB : EF = EF : CD. In Book VI, Proposition 13, Euclid shows that EF can be found by constructing a semi-circle of diameter ABD (connecting point C to B), and draw a perpendicular line from B to the arc of the semi-circle at G. Then BG is the required line. Note that there are many mean proportions in dividing a line AD into 2 parts.

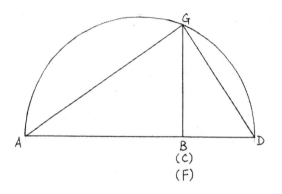

Figure 2.7 <u>Mean Proportion of 2 Lines</u>

Proof

From Thales' theorem (section 2.1.8), ∠AGD is 90°, and also ∠GCD is 90°.

Let	∠BAG = α,
then	∠BGA = 90° - α
so	∠CGD = α

Therefore ΔABG and ΔGCD are similar as all 3 interior angles are equal respectively

Hence	AB : BG = CG : CD
or	AB : BG = BG : CD

2.2.2 <u>Book IV, Proposition 14</u>

In [two] equal [in area] and equiangular (equal in angles) parallelograms, the sides about the equal angles are reciprocally

proportional; and equiangular parallelograms in which the sides about the equal angles are reciprocally proportional are equal [in area].

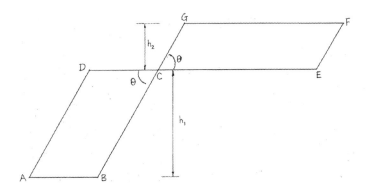

Figure 2.9 <u>Illustration for Proposition 14,</u>
<u>Book 4 of *The Elements*</u>

Proof

Let ABCD and CEFG be equal and equiangular parallelograms having the angles in C equal.

Let BC and CG be connected in a straight line. Then DC and CE are also in a straight line.

Area of parallelogram ABCD $= DC \times h_1$
$$= DC \times BC\sin\theta$$

Area CEFG $= CE \times h_2$
$$= CE \times CG\sin\theta$$

Since the 2 areas equal, $DC \times BC\sin\theta = CE \times CG\sin\theta$

or $\qquad \dfrac{DC}{CE} = \dfrac{CG}{BC}$

i.e. the sides are reciprocally proportional about the equal angles.

The reverse is also true since if 2 parallelograms with an equal angle have sides reciprocally proportional about the equal angles, then by adding $\sin\theta$ (θ being the equal angle) on both side of the equation, their areas are shown to be equal.

Note that the proof is a simplified version of Euclid's original one.

2.2.3 Book IV, Proposition 30

To cut a given finite straight line in extreme and mean ratio.

Let us divide a straight line AB at a point C lying between A and B, then the 2 segments (part of a line) AC and CB is in the ratio AC : CB. If the ratios AB : AC and AC : CB equal to each other, then the line AB is said to have been divided in the **Extreme Proportion** (at C). Note that the extreme proportion is one of the mean proportions, and AC > (greater than) CB.

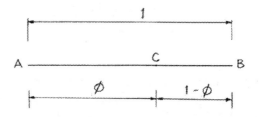

Figure 2.10 Extreme Proportion
(Note that here $\phi = 0.618$)

In Book VI, Proposition 30, a method is proposed to cut (divide) a straight line in extreme and mean ratio. This is a graphical or geometric method, the technique is by trial and error to find two equal areas. It is typical that in ancient times where algebraic theories were not well established, algebraic problems were sometimes solved by geometric means. Better methods of dividing the line in the golden ratio will be discussed later.

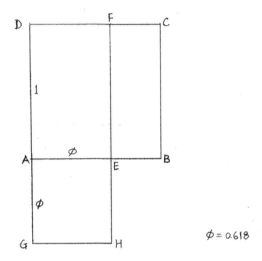

Figure 2.11 <u>Illustration for Proposition 30,</u>
<u>Book 4 of *The Elements*</u>

The following is an interpretation of the proposition. Referring to Fig 2.11, AB is the finite line and it is required to cut it in extreme and mean ratio. First form a square ABCD with side equals to AB. Then construct a parallelogram (rectangle) DGHF such that AGHE is a square and its area equals to the smaller rectangle EBCF. (In other words, area of rectangle DGHF equals area of square ABCD). Then E is the extreme and mean ratio.

Here is a simple informal proof. Assume AB = 1 and the proposition holds. Then AE = ϕ (0.618) and AG = 0.618.

Hence   DG   = DA + AG

= 1 + 0.618 = 1.618

Area   DGHF = DG x AE

= 1.618 x 0.618

= 1,   area of the square ABCD.

The formal proof of this proposition by Euclid is as follows:

Let AB be the given finite straight line and the square ABCD be constructed. Extend the edge DA to G and form the square AGHE, similar but smaller than ABCD. Since [area] DGHF equals to ABCD (as required), let AEFD be subtracted from each. Then AGHE equals EBCF, and is also equiangular with it. According to Proposition 14, the sides of the two equal [area] parallelograms about the equal angles are reciprocally proportional.

$$\frac{HE}{EF} = \frac{EB}{AE}$$

But HE = AE, EF = AB,

i.e. $\frac{AE}{AB} = \frac{EB}{AE}$ which is the ratio of greater segment to the whole

equals to the smaller segment to the greater segment, the mean and extreme proportion.

Now let us try to find the ratio by trial and error. It is required to find AE such that

$$AE^2 = EB \times EF$$

Let AB = 1, EF = AB = 1, EB = 1 – AE

i.e. to find $\qquad AE^2 = EB$

Try AE = 0.6, AE^2 = 0.36, EB = 0.4, EB > AE^2.

Try AE = 0.63, AE^2 = 0.3969, EB = 0.37, AE^2 > EB.

Try AE = 0.62, AE^2 = 0.3844, EB = 0.38

So ϕ = 0.62 up to 2 decimal places.

In fact, the proposition is just a restatement of the definition of the golden ratio in geometric form.

2.2.4 Book XIII, Proposition 1

If a straight line be cut in extreme and mean ratio, the square on the greater segment added to the half of the whole, is five times the square on the half.

First we define a term used by Euclid, known as *gnomon*.

A *gnomon* is any figure that when added to another figure (original one), gives the resultant figure similar to the latter (original). This represents successive increments of growth in the original figure.

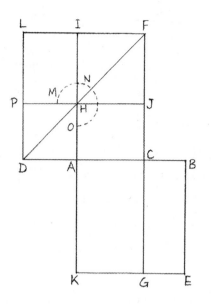

Figure 2.12 <u>Illustration for Proposition 1,</u>
<u>Book 13 of *The Elements*</u>

Proof

Let the straight line AB be cut in extreme and mean ratio at point C, and AC be the greater segment. Extend BA to point D such that AD is half of AB. Form the square CDLF (square of (AC+ AB/2)), and the smaller square ADPH (square of AB/2).

Construct the square ABEK (square of AB) and extend the line FC to point G. Also extend AH to I and PH to J.

Since AB has been cut in extreme and mean ratio at C, area of rectangle CBEG equals to square HIFJ (square of AC) (refer to Section 2.2.3).

Now AH = AD, AK = AB, and AB = 2 x AD,

Therefore AK = 2 x AH

and area of rectangle ACGK = 2 x AHJC (with common edge AC)

But rectangle HPLI = AHJC,

therefore ACGK = AHJC + HPLI

But it has been shown that CBEG = HIFJ,

Hence the whole square ABEK = gnomon MNO,

which equals to 4 times square ADPH.

Adding gnomon MNO to ADPH gives the square CDLF, therefore square CDLF equals to 5 x square ADPH.

Note that the proposition when written in algebraic form, gives one of the irrational value of ϕ.

For let AB = 1,

$$(\phi + \tfrac{1}{2})^2 = 5(\tfrac{1}{2})^2$$
$$\phi + \tfrac{1}{2} = \sqrt{5}(\tfrac{1}{2})$$
$$\phi = \frac{\sqrt{5}-1}{2}$$

2.2.5 Book XIII, Proposition 4

If a straight line be cut in extreme and mean ratio, the square on the whole and the square on the lesser segment together are triple of the square on the greater segment.

Proof

Let AB be the straight line and cut in extreme and mean ratio at C, with AC being the greater segment. Form the large square ABED and the two smaller squares HFGD (greater segment) and CBKF(lesser segment).

It can be seen that the rectangles ABKH and CBEG are equal. Since AB is cut in extreme and mean ratio at C, area HFGD (square of AC) equals to rectangle CBEG, and also equals to rectangle ABKH. (refer to Section 2.2.3)

Area ABED equals to square HFGD + gnomon LMN, hence equals to 3 times HFGD less the extra square CBKF.

In other words, $$AB^2 = 3AC^2 - CB^2$$
$$AB^2 + CB^2 = 3AC^2$$

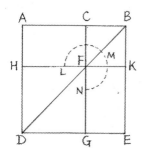

Figure 2.13 <u>Illustration for Proposition 4,</u>
<u>Book 13 of *The Elements*</u>

2.2.6   <u>Book IV, Proposition 5</u>

If a straight line be cut in extreme and mean ratio, and there be added to it a straight line equals to the greater segment, then the whole [extended] line has been cut in extreme and mean ratio, and the original line is the greater segment.

Proof

Let the straight line AB be cut in extreme and mean ratio at point C, and AC be the greater segment. Extend the line to D with AD = AC.

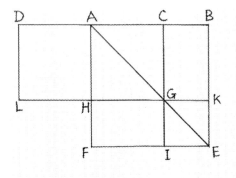

Figure 2.14 <u>Illustration for Proposition 5,</u>
<u>Book 13 of *The Elements*</u>

Form squares ABEF (with side AB) and ACGH (side AC).

Since AB is cut in extreme and mean ratio at C, by Section 2.2.3, area of square ACGH equals to rectangle CBEI.

But square DAHL = ACGH, and rectangle HKEF equals to CBEI (since CBKG = HGIF), hence

Rectangle HKEF = Square DAHL

adding ABKH to both side,   ABEF = DBKL

Therefore the larger square ABEF equals to the larger rectangle DBKL. By Section 2.2.3, line DB has been cut in extreme and mean ratio at A, and AB is the greater segment.

Note that this proposition is important in showing that the golden ratio gives rise to fractal geometry.

2.2.7   Book XIII, Proposition 8

In an equilateral and equiangular pentagon, straight lines subtend two angles (diagonal lines) *taken in order* (adjacent to one another), *they cut one another in extreme and mean ratio, and their greater segments are equal to the side of the pentagon.*

Consider an equilateral and equiangular (i.e. regular) pentagon ABCDE. Let the diagonals AC and BE cutting each other (intercept) at H. Then the two lines are cut at extreme and mean ratio at H, and the greater segments CH and EH equal to the side of the pentagon.

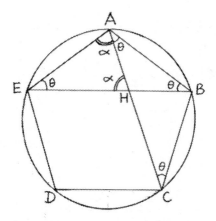

Figure 2.15 Illustration for Proposition 8, Book 13 of *The Element*

Proof (simplified)

Construct the circumscribed (i.e. touching all the vertices) circle of the pentagon. Consider $\triangle$ ABE and let $\angle$ABE = θ.

Since AB = AE (edge of regular pentagon)

$\angle$AEB = $\angle$ABE = θ

Similarly for $\triangle$ ABC, AB = BC and $\angle$BAC = $\angle$BCA = θ.

Therefore both $\triangle$s ABE and ABC are isosceles and they are equal.

Let $\angle$AHE = α,

In $\triangle$ ABH, $\alpha = 2\theta$ (exterior angle at H).

Consider the arcs BC and CDE, CDE is twice that of BC, therefore $\angle$EAC = 2 $\angle$BAC = 2θ = α

And $\triangle$ AEH is isosceles, AE = EH.

Since $\triangle$ABE and $\triangle$ HAB have 2 angles equal, the vertices must also be equal and the two $\triangle$s are similar.

Thus $\quad \dfrac{AB}{BE} = \dfrac{BH}{AB}$

and $\quad$ AB = AE = EH

Hence $\quad \dfrac{EH}{BE} = \dfrac{BH}{EH}$

Therefore diagonal BE has been cut in extreme and mean ratio at H, and the greater segment EH equals to the side of the pentagon AE.

By symmetry, diagonal AC has also been cut in extreme and mean ratio at H and CH is the greater segment.

Let us go one step further from this proposition. Join the diagonal AD meeting EH at J. Since all three angles are equal, Δs ACD and AHJ ae similar.

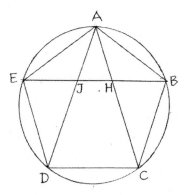

Figure 2.16 <u>Illustration for Proposition 8 (extension),</u>
<u>Book 13 of *The Elements*</u>

Hence $\dfrac{JH}{AJ} = \dfrac{DC}{AD}$

But $AD = BE$, $DC = EH$ (EH = side of pentagon)

or $\dfrac{JH}{AJ} = \dfrac{EH}{BE}$

As $\dfrac{EH}{BE} = \dfrac{HB}{EH}$, ($BE$ being cut in golden ratio at H)

Hence $\dfrac{JH}{AJ} = \dfrac{HB}{EH}$

And $AJ = EJ$, $HB = EJ$ by symmetry,

Therefore $\dfrac{JH}{EJ} = \dfrac{EJ}{EH}$

i.e. EH is cut at extreme and mean ratio at J.

This shows that the regular pentagon can give rise to fractal geometry, which will be discussed later.

2.2.8 Book XIII, Proposition 9

If the side of the [regular] hexagon and that of the [regular] decagon inscribed in the same circle be added together, the whole [resultant] line has been cut in extreme and mean ratio, and its greater segment is the side of the hexagon.

In other words the proposition states that the sides of a regular hexagon and regular decagon inscribed in the same circle is in golden ratio, and the sides added together is also in golden ratio to that of the hexagon alone.

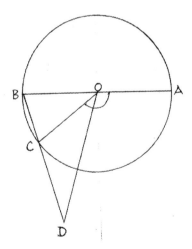

Figure 2.17 Illustration for Proposition 9,
Book 13 of *The Elements*

Proof

Let ABC be a circle with centre O, and BC be a side of a decagon. Extend the line BC to D, CD being a side of the hexagon inscribed in the same circle.

Join OB, OC and OD, and extend BO to A (AB is diameter of circle ABC).

Since   BC is a side of a regular decagon, the arc of the semi-circle ACB is 5 times the arc CB, therefore arc AC is quadruple (4 times) arc CB.

Hence $\angle AOC = 4 \angle COB$.

Since $\angle OBC = \angle OCB$ (isosceles $\triangle$), and these 2 angles add up equals to $\angle AOC$ (exterior angle), therefore

$\angle AOC = 2 \angle OCB$.

Now OC = CD (side of a regular hexagon equals to the radius of the inscribed circle)

Thus $\triangle COD$ is isosceles, $\angle COD = \angle CDO$,

And $\angle OCB$ is twice $\angle COD$ (exterior angle)

But $\angle AOC = 2 \angle OCB$,

Hence $\angle AOC = 4 \angle COD$, and as also $\angle AOC = 4 \angle COB$,

Therefore $\angle COD = \angle COB$.

$$\angle BOD = \angle COB + \angle COD = 2 \angle COD$$

$$= \angle OCB = \angle OBC,$$

This implies $\triangle DBO$ is isosceles and similar to $\triangle OBC$

Hence $\dfrac{BC}{CD} = \dfrac{CD}{BD}$

But $\qquad$ OB = CD

Therefore $\qquad$ $\dfrac{BC}{OB} = \dfrac{OB}{BD}$

That is, BD is cut at extreme and mean ratio at C, and CD is the greater segment.

Chapter 3

Golden Ratio Demystified through Elementary Algebra and Trigonometry

3.1   Algebraic Terms

An algebraic equation contains one or more variables or unknowns, usually denoted by x, y, z etc. The purpose is to solve (find out) the unknowns. In general if there are n unknowns, we need n equations or governing conditions in order to determine all the unknowns independently. Besides unknowns, there are constants. If a constant is attached to (i.e. multiply with) an unknown term, the constant is called the *coefficient* of the term. For example in the following equation with 2 unknowns x and y:

$$ax + bxy + cy + d = 0$$

a, b, c are coefficients and d is the constant term.

We will discuss only equations with one unknown x. If the highest power of x is 1, the equation is called 1st order or linear. If the highest power of x is 2, then 2nd order or quadratic, and so forth.

So $\quad$ $ax + b = 0$ $\qquad$ is 1st order or linear

$\qquad$ $ax^2 + bx + c = 0$ $\qquad$ is 2nd order or quadratic

$\qquad$ $ax^3 + bx^2 + cx + d = 0$ $\quad$ is 3rd order

3.2 $\quad$ Fundamentals of Quadratic Equation

Let us review some fundamental formulae for solving quadratic equations.

3.2.1 $\quad$ Solving $ax^2 + bx + c = 0$

Consider linear equation with one unknown, the solution can be obtained immediately.

For $\quad$ $ax + b = 0$

then $\qquad$ $x = - b /a$

First review the formula for $(p + q)^2$

By simple multiplication,

$(p + q)^2 = (p + q)(p + q) = p^2 + pq + pq + q^2$, $\qquad$ hence

$(p + q)^2 = p^2 + 2pq + q^2$ ....(1)

We can solve quadratic equation by the method of "completing the square".

Let $\qquad$ $ax^2 + bx + c = 0$ $\qquad$ where a, b, c are constants

Divided by a, $\quad$ $x^2 + (b/a)x = - c/a$

To complete the square, add $(b/2a)^2$ on both sides

Thus $(x + b/2a)^2 = (b/2a)^2 - c/a$

$$x + b/2a = \sqrt{\frac{b^2 - 4ac}{4a^2}}$$

Therefore
$$x = \frac{-b \pm \sqrt{b^2 - 4ac}}{2a} \quad \ldots\ldots\ldots..(2)$$

Let $b^2 - 4ac = \Delta$, known as the *determinant*

Note that there are 2 solutions or *roots* for x, namely

$x_1 = (-b + \sqrt{\Delta}) / 2a$,

$x_2 = (-b - \sqrt{\Delta}) / 2a$

If Δ is negative, i.e. $4ac > b^2$,

then there is no real solution for x.

3.2.2   Properties of the Roots of Quadratic Equation

Let α, β be the roots of the quadratic equation

$$ax^2 + bx + c = 0$$

Then $x = \alpha$ or $x = \beta$

Or $(x - \alpha)(x - \beta) = 0$

$$x^2 - (\alpha+\beta)x + \alpha\beta = 0$$

Divide the original equation by a,

$$x^2 + (b/a)x + c/a = 0$$

Compare the last 2 equations,

Hence $\qquad \alpha + \beta = -b/a$ ....(3)

$\qquad\qquad\quad \alpha\beta = c/a$ ....(4)

3.3 Algebraic Solutions of the Golden Ratio

3.3.1 Golden Ratio as an Irrational Number

Going back to the definition of the golden ratio, a line AB is divided in the golden ratio at C if the proportion AC to AB equals to that of CB to AC.

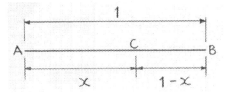

Without loss of generality, let $AB = 1$

and x = length or proportion of the greater segment

Then $\qquad \dfrac{x}{1} = \dfrac{1-x}{x}$

or $\qquad\qquad x^2 = 1 - x$

$\qquad\qquad x^2 + x - 1 = 0$ ....(5)

This is a quadratic equation with $a = 1$, $b = 1$, $c = -1$

Apply equation (2),

the solution is $\qquad x = \dfrac{-b \pm \sqrt{b^2 - 4ac}}{2a}$

$$= \frac{-1 \pm \sqrt{1-(-4)}}{2}$$

Therefore $\qquad \phi = \frac{\sqrt{5}+1}{2} \cong 0.618$

The negative root being neglected.

Put it in another way: let the greater segment AC be x as before, but the smaller segment CB be 1 (the ratio is x). The overall length AB is thus x + 1. Here x > 1.

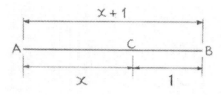

Then by definition of the golden ratio,

$$\frac{1}{x} = \frac{x}{x+1}$$

Hence $\qquad x^2 = x + 1$

$$x^2 - x - 1 = 0 \quad\ldots\ldots\ldots\ldots\ldots\ldots\ldots\ldots\ldots\ldots\ldots(6)$$

This gives $\qquad x = \frac{1 \pm \sqrt{5}}{2}$

or $\qquad \phi = \frac{\sqrt{5}+1}{2} \cong 1.618$

The negative root being neglected.

Here we understand that there are 2 values of ϕ, one > 1 and the other < 1, and they are derived from 2 different equations.

Let $\phi_1 > 1$, $\phi_2 < 1$

Then
$$\frac{1}{\phi_1} = \frac{2}{\sqrt{5}+1}$$

$$= \frac{2(\sqrt{5}-1)}{(\sqrt{5}+1)(\sqrt{5}-1)}$$

$$= \frac{2(\sqrt{5}-1)}{5-1}$$

$$= \frac{\sqrt{5}-1}{2} = \phi_2$$

Note that from now on ϕ without the subscript means ϕ_1

3.3.2 Solving the Pyramid Problem

Let us go back to Section 1.2, where John Herschel translated the ancient Greek Herodotus's words that "each face of a [square] pyramid equals to the square of its height".

We shall see that this lead to a golden ratio.

Let   2a = one side of the square base pyramid
         b = altitude of the triangular face
         h = height of the pyramid

Then area of one face $= \frac{1}{2}(2a)b = ab$

By Herodotus, $ab = h^2$

$\qquad\qquad\qquad = b^2 - a^2$ (Pythagoras' Theorem, refer to next section)

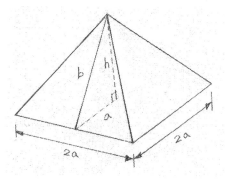

Divided by a^2 $(b/a) = (b/a)^2 - 1$

Or $(b/a)^2 - (b/a) - 1 = 0$

Let $b/a = x$, $x^2 - x - 1 = 0$

This is the equation of the golden ratio, equation (6)

And the solution is $x = \phi$

Or altitude of the face (apothem) of the pyramid to half the base is in golden ratio.

3.4   Pythagoras' Theorem

The theorem is attributed to the ancient Greek mathematician Pythagoras (570 -495 BC) for his proof of the theorem, although some archaeologists discovered that the Babylonians had recorded the Pythagorean triples (a, b, c all natural numbers or positive integers) over a thousand years before Pythagoras' time.

The Pythagoras' or Pythagorean Theorem states that in a right angle triangle, the sum of squares of the sides sustain the right angle equals to the square of the hypotenuse (side opposite to the right angle).

i.e. $a^2 + b^2 = c^2$

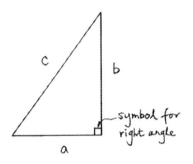

3.4.1 Proof of the theorem

There are numerous proofs of the Pythagoras Theorem worldwide from ancient time up to present. We will only discuss two, one concerning area and one by similar triangles.

(1) Construct 2 squares $(a + b)^2$ with configurations as below:

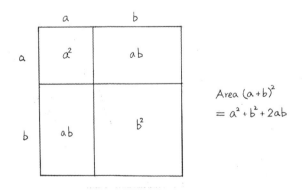

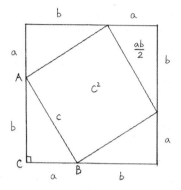

The areas of the above 2 squares are both equal.

Therefore $\quad a^2 + b^2 + 2ab = c^2 + 4(ab/2)$

Hence $\qquad a^2 + b^2 = c^2$

(2)   Construct a right angle triangle ABC. Join a line CD, perpendicular to AB, the hypotenuse side. Then both triangles BCD and ACD are right angle.

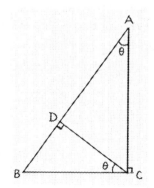

Let $\angle BAC = \theta$,

In Δs ABC and CBD, $\angle ABC$ is common and both have a right angle; therefore $\angle BCD = \theta$ and the 2 Δs are similar.

Hence $\qquad \dfrac{BC}{AB} = \dfrac{BD}{BC}$

or $\qquad BC^2 = AB \times BD$

Similarly, Δs ABC and ACD are similar.

Hence $\qquad \dfrac{AC}{AB} = \dfrac{AD}{AC}$

or $\qquad\qquad AC^2 = AB \times AD$

Therefore $\quad BC^2 + AC^2 = AB \times BD + AB \times AD$

$$= AB \, (BD + AD) = AB \times AB$$

$$= AB^2$$

Or $\qquad\qquad a^2 + b^2 = c^2 \quad$ where $BC = a$, $AC = b$, $AB = c$

Note that for $a = 3$, $b = 4$, $c = 5$, the sides are in arithmetic progression ($b = a + 1$, $c = b + 1$), and is the only right angle triangle which sides are in arithmetic progression.

For $a = 1$, $b = \sqrt{\phi}$, $c = \phi$, the sides are in geometric progression ($b = \sqrt{\phi}\, a$, $c = \sqrt{\phi}\, b$), and is the only right angle triangle in geometric progression. This triangle is also known as the *Kepler Triangle*.

3.4.2   The Egyptian Triangle

Ancient Egyptian engineers used the 3, 4, 5 triangle to establish the right angle or "square" boundary of the farm land. So it is known as the Egyptian triangle. Also, the triangle is used to symbolise the

Egyptian trinity: 3 = Osiris, 4 = Isis, and 5 = Horis. By bisecting the angle at the base, an interesting relation can be seen.

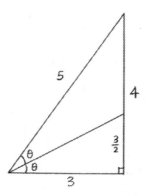

Figure 3.1 <u>Bisecting the 3, 4, 5 triangle</u>

Here the compound angle formula for tangent has to be introduced. Those who are not familiar with the formula or could not remember it just take it for granted.

Recall $\qquad \tan 2\theta = \dfrac{2tan\theta}{1 - tan^2\theta}$

$\qquad\qquad\qquad = \dfrac{4}{3}$ $\qquad$ (from fig 3.1)

Hence $\qquad 2- 2tan^2\theta = 3tan\theta$

$\qquad\qquad 2tan^2\theta + 3tan\theta -2 = 0$

$\qquad\qquad (2tan\theta - 1)(tan\theta + 2) = 0$

Thus $\qquad 2tan\theta - 1 = 0$

And $\qquad tan\theta = \frac{1}{2}$

or $\qquad tan\theta + 2 = 0 \qquad$ discard this negative root.

Hence θ = arc tan $\frac{1}{2}$

This is equal to the apex angle of the 1:2 triangle. That means the remaining right angle triangle is 1:2 !

3.4.3 Pythagorean Triples

One obvious question about the Pythagoras' Theorem is can all a, b, c be whole numbers? The answer is yes, and the most quoted one is 3, 4, 5 or their multiples $(3^2 + 4^2 = 5^2)$. Then is there a general rule for finding a, b, c all in whole or natural numbers? The answer is also yes, and such set of numbers are known as *Pythagorean Triples*.

For $\qquad a^2 + b^2 = c^2$

The formula is $\underline{a = m^2 - n^2}$

$\qquad\qquad\quad \underline{b = 2mn}$

$\qquad\qquad\quad \underline{c = m^2 + n^2}$

where m, n are natural numbers

While the proof is rather tedious and will not be covered here, there are certain assumptions or deductions that have to be observed.

1) a, b, c are relative prime, i.e. no common factors. They are referred as <u>primitive</u> Pythagorean Triple.

2) a and b cannot be both odd or both even, it has to be one odd and one even.

3) Following (2) above, c must be odd.

4) m > n, also m and n cannot be both odd or both even, otherwise the triple obtained is not primitive.

First start with m = 2, n = 1. Then a = 3, b = 4, c = 5.

Next try m = 3, n = 2. We obtain a = 5, b =12, c = 13.

Let us determine the characteristics of triples in which

$$m = n + 1,$$

Then $a = m^2 - n^2 = (n + 1)^2 - n^2 = (n^2 + 2n + 1) - n^2$

$$= 2n + 1$$

$$b = 2mn = 2(n + 1)n = 2n^2 + 2n$$

And $c = m^2 + n^2 = (n + 1)^2 + n^2 = (n^2 + 2n + 1) + n^2$

$$= (2n^2 + 2n) + 1$$

Or $c = b + 1$

If $m = n + 3$, then $c = b + 9$ (try it yourself)

In general if $m = n + p$,   p being any natural number,

then $c = b + p^2$

The following table listed out the first 8 and a few other primitive Pythagorean Triples.

m	n	a	b	c
2	1	3	4	5
3	2	5	12	13
4	1	15	8	17
4	3	7	24	25
5	2	21	20	29
6	1	35	12	37
6	5	11	60	61
7	2	45	28	53
8	5	39	80	89
13	8	105	208	233
34	21	715	1428	1597

Table 3.1 <u>Pythagorean Triples</u>

3.5   A Very Simple Way to Construct the Golden Section

It has been found that many ancient civilizations had used the golden ratio in their architecture and construction works. I would propose that the application of the ratio is by chance or accidental, and by drawing a right angle triangle with base to height of 1 : 2, the golden rectangle can be obtained easily and hence the ratio. The other tool needed is a pair of compasses, which might be just a string attached to a peg at one end, that ancient peoples were quite familiar with.

3.5.1   Golden Rectangle

Here is how it works. Draw a right angle triangle OAB with height AB double the base OA. Let OA = 1 unit, then OB = $\sqrt{5}$.

Construct a semi-circle with O as centre, OB as radius, meeting the extended baseline OA at C and D.

Then   AC = $\sqrt{5} + 1$,   AD = $\sqrt{5} - 1$

Hence   AC : AB = $\sqrt{5} + 1 : 2 = \phi_1$

And   AD : AB = $\sqrt{5} - 1 : 2 = \phi_2$

Therefore 2 golden rectangles are obtained,

i.e. ABEC and ADFB.

Is the golden ratio really so mysterious?

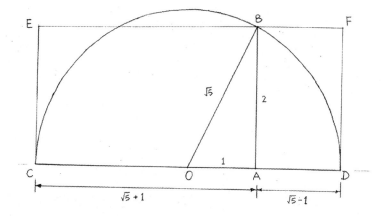

Figure 3.2 <u>Construction of the Golden Rectangle from a Right Angle Triangle</u>

3.5.2   <u>Mean and Extreme Proportion</u>

Applying the 1 : 2 triangle, it is quite easy to divide a line in mean and extreme proportion by geometric construction.

Let AB be the line to be divided. Draw a perpendicular line from A and extend the horizontal line also from A. Use a pair of compasses and A as centre, transfer the length AB to point C on the perpendicular.

Next bisect AB at point M. Let AB = 1, then AM = ½. Take A as centre, transfer length AM to AD by the compasses. Hence triangle ACD is 1 : 2.

Take D as centre, DC as radius, swing the arc CE meeting line AB.

Then $AE = DE - DA = \dfrac{\sqrt{5}}{2} - \dfrac{1}{2}$

$\qquad = \phi_2$

Therefore E divides AB in the mean and extreme ratio.

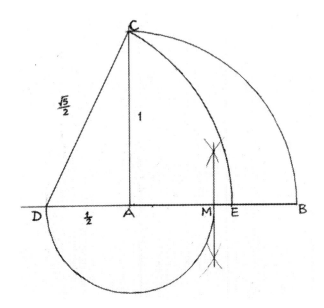

Figure 3.3 <u>Divide a Line in the Golden Ratio</u>
<u>from a 1:2 Right Angle Triangle</u>

Chapter 4

Geometry and Trigonometry Related

4.1   Triangles in Golden Ratio

4.1.1   <u>Golden Triangle</u>

Let us construct a few isosceles triangles with ratio ϕ : 1.

First let the base be ϕ unit and height 1 unit. The angle between the base and the hypotenuse (θ) is

arc tan($1/0.5\phi$) or 51.03°.

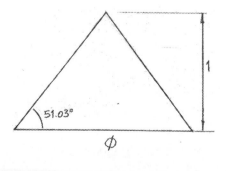

Let base be 1 and height ϕ. Then θ = arc tan $(\phi/\frac{1}{2})$ = 72.83°.

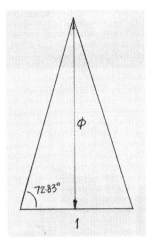

Let base be 1 and hypotenuse be ϕ. (Fig 4.1)

Then θ = arc cos $(\frac{1}{2}/\phi)$= 72° exactly!

Hence $\triangle$ABC is taken as the *Golden Triangle*.

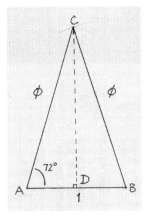

Figure 4.1 <u>The Golden Triangle</u>

4.1.2 Golden Gnomon

Consider the triangle with ϕ as base and 1 as hypotenuse.

θ = arc cos (1 / $\frac{1}{2}\phi$)

  = 36°,   also an exact angle

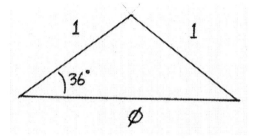

Figure 4.2 <u>The Golden Gnomon</u>

This triangle is known as the *Golden Gnomon* (we shall see why), or sometimes as the golden triangle of the 2nd kind [obtuse (apex angle > 90°) triangle].

Note that angles in the golden triangle is in the proportion of 1 : 2 : 2, and that in the golden gnomon is 3 : 1 : 1. Both are unique, i.e. no other triangles have these proportions.

Let us bisect one of the side angles of the golden triangle ABC. Then ∠DAB = 36°, ∠ABD = 72°, thus ∠ADB = 72°.

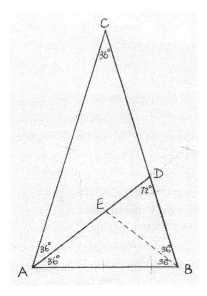

Hence Δ ABD is also a golden triangle. In Δ ADC, ∠ACD = ∠DAC = 36°, implies ∠ADC = 108°. According to the definition of gnomon in section 2.2.4, Δ ADC is a gnomon, and hence the golden gnomon. Also, it can be shown that the ratio of Area ADC to Area ABD is φ. (Hint area of Δ is ½ height x base.)

Therefore by bisecting one base angle continuously, the golden triangle exhibits fractal geometry.

Note that the golden gnomon also shows fractal property if a golden triangle is taken away from the original form.

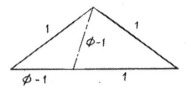

From fig.4.1, $\quad \cos 72° = \dfrac{½}{\phi}$

$$= \dfrac{1}{\sqrt{5}+1} = \dfrac{\sqrt{5}-1}{\left(\sqrt{5}+1\right)\left(\sqrt{5}-1\right)}$$

$$= \dfrac{\sqrt{5}-1}{4}$$

And from fig. 4.2, $\quad \cos 36° = \dfrac{\phi/2}{1}$

$$= \dfrac{\sqrt{5}+1}{4}$$

$$\cos^2 72° + \cos^2 36° = \dfrac{\left(\sqrt{5}-1\right)^2 + \left(\sqrt{5}+1\right)^2}{4^2}$$

$$= \dfrac{5 - 2\sqrt{5} + 1 + 5 + 2\sqrt{5} + 1}{16} = \dfrac{12}{16}$$

$$= \dfrac{3}{4} = (\sqrt{3}/2)^2 = \cos^2 30° \ !$$

Hence if we construct a right angle triangle with cos 72° and cos 36° as the base and height respectively, the hypotenuse will be cos 30°. The angle of this triangle is

arc tan $[(\sqrt{5}+1) / (\sqrt{5}-1)]$ = arc tan $[(3+\sqrt{5})/2]$

$$= \text{arc tan } (1+\phi) = \text{arc tan } \phi^2$$

$$= 69.094°$$

Since $\quad \sin\theta = \cos(90° - \theta)$

Hence $\quad \sin^2 18° + \sin^2 54° = \sin^2 60°$

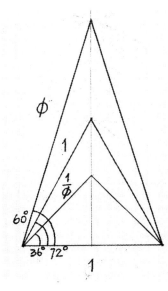
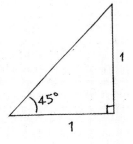

Figure 4.3 <u>Triangles in which the angles produce</u>
<u>exact trigonometric values</u>

4.1.3   <u>Bisecting the Angle of the Golden Triangle Successively</u>

The side angle of the golden triangle can be bisected successively in 4 ways, each will form different graphics by connecting the vertices of the triangles.

1) Wholly clockwise

2) Wholly counter-clockwise

3) Alternatively, clockwise first

4) Alternatively, counter-clockwise first.

Here clockwise means the bisected angle is on the right side of the apex of the triangle, and counter-clockwise on the left side.

The following graphs are obtained.

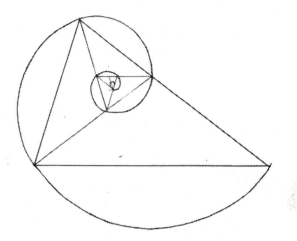

Figure 4.4.1 <u>Golden Triangle successively bisected</u>
<u>Clockwise</u>

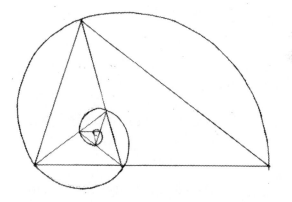

Figure 4.4.2 <u>Golden Triangle successively bisected</u>
<u>counter-clockwise</u>

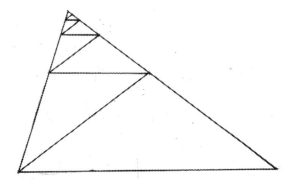

Figure 4.4.3 <u>Golden Triangle successively bisected alternatively, clockwise first</u>

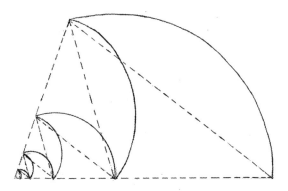

Figure 4.4.4 <u>Golden Triangle successively bisected alternatively, counter-clockwise first</u>

The first 2 graphs show the construction of the logarithmic spiral, or approximates to the golden spiral. The first one expands counter-clockwise and the second clockwise. Curves are obtained by joining the base vertices of the gnomons obtained after bisecting the angle. The third figure exhibits fractal nature of the reduced triangles.

Connecting the base points of the gnomon by an arc with apex as centre, the last one is a form of wave shape, also in fractal geometry.

4.1.4   Golden Rhombus

A golden rhombus is one with width and height in the ratio ϕ :1.

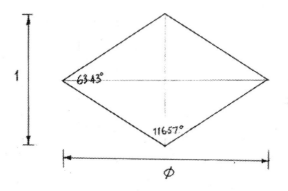

Figure 4.5 The Golden Rhombus

Thus the rhombus formed by 2 golden triangles or 2 golden gnomons is not a golden rhombus. The golden rhombus can be found in 3 dimensional polyhedrons.

4.1.5   Kepler Triangle

From the equation $1 + \phi = \phi^2$,

We can form a right angle triangle with the triple 1, $\sqrt{\phi}$, ϕ.

This is known as Kepler Triangle, after Kepler who first described it. The edges of the triangle are in geometric progression, the ratio being $\sqrt{\phi}$. One interesting finding is that by constructing a circle of diameter ϕ, being the hypotenuse of the triangle, and a square of side $\sqrt{\phi}$, being the longer edge of the triangle, the relationship between π and ϕ can be determined.

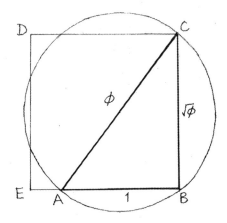

Perimeter of square $= 4\sqrt{\phi} = 5.0881$

Perimeter of circle $= \pi\phi = 5.0832$

The difference $= 0.096\%$

Therefore $\qquad 4\sqrt{\phi} \approx \pi\phi$

Or $\qquad\qquad \pi \approx \dfrac{4}{\sqrt{\phi}}$ or $4\sqrt{\phi}_2$

4.2   Pentagon and Pentagram

Unless expressed otherwise, all polygons mentioned are in regular form, that is, all sides of the polygon are equal. By connecting the diagonal lines of the pentagon, the pentagram is obtained. Its five pointed edges are all golden triangles (Sect 2.2.7). Pentagram is a common symbol in different religions and national emblems over the world.

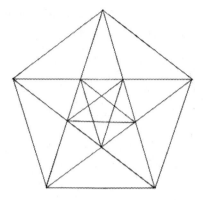

Figure 4.6 <u>Pentagon and Pentagram</u>

Referring to fig 4.6, it can be seen that the core of the pentagram is a pentagon. In the other way, five golden triangles added externally to the edges of the pentagon produce the pentagram. Hence both pentagon and pentagram are fractal.

Pentagon (and hence pentagram) can be constructed inside a base circle (inscribed) or touching the circle (circumscribed) graphically by the 1:2 right angle triangle. One may say that any polygon can easily be drawn using a protractor! The point is that in (classical) geometry, the only tools allowed are straight edge and compasses. Also, the figure obtained by geometric method should be more accurate and beautiful.

4.2.1 <u>Construct a Pentagon inscribed in a Circle</u>

In fig 4.7, let O be the centre of the base circle and the radius R = 1 unit. Construct the right angle triangle OCE where E is obtained by bisecting OB. Hence △ OCE is in 1:2, and

CE= √5/2. We have to find the distance from O to the base of the pentagon, which is Rcos36°. Remember that R=1 and

cos36° = (√5+1)/4, the aim is to produce a perpendicular from O with the value of cos36° (ϕ/2).

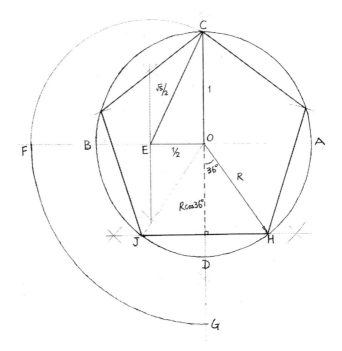

Figure 4.7 <u>Constructing the Pentagon</u>
<u>inscribed in a circle</u>

Use E as centre, swing the radius EC counter-clockwise to the horizontal line AB at F. Now OF= OE + EF = ½ + √5/2 = ϕ, take O as centre swing F to G. Then bisect OG and obtain the bisect line JH touching the circle. The distance from O to JH is ϕ/2. JH is the base edge of the pentagon. Take JH as the radius and get the other vertices along the circle to complete the figure.

4.2.2   Construct a Pentagon circumscribes about a Circle

From the circumscribed circle we find the inscribe circle. Let OP be radius of the inscribe circle, and R radius of the circumscribed circle be 1 unit. Then OP equals to R / cos36° or 1 / cos36°.

$$1 / \cos36° = \frac{4}{\sqrt{5}+1} = \frac{4\left(\sqrt{5}-1\right)}{\left(\sqrt{5}+1\right)\left(\sqrt{5}-1\right)}$$

$$= \frac{4\left(\sqrt{5}-1\right)}{\left(5-1\right)} = \sqrt{5} - 1$$

Therefore   OP = ($\sqrt{5}$ – 1) R

Form the 1:2 right angle triangle OEC by bisecting the radius OB to obtain E, OC being radius also (fig 4.8). Take E as centre and EC as radius, swing an arc clockwise to meet the horizontal line AB at F. Then OF equals to ($\sqrt{5}$ – 1)/ 2. Use F as centre swing the length OF to the other side of the horizontal to get G. OG being double of OF, is the radius of the inscribe circle. Draw the circle, and from P, construct lines tangent (touching one point) to the circumscribed (inner) circle and end at the inscribe (outer) circle to obtain PQ and PS. At D, construct horizontal tangent line touching the outer circle at both ends. These are edges of the pentagon. Complete the remaining edges.

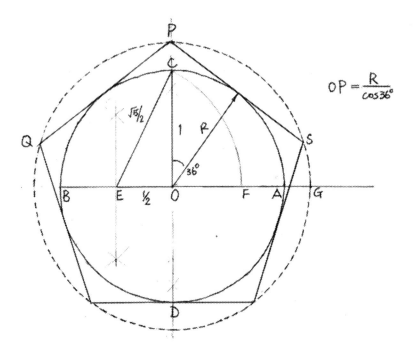

$$OP = \frac{R}{\cos 36°}$$

Figure 4.8 <u>Constructing the Pentagon</u>
<u>Circumscribes about a Circle</u>

4.3 Hexagon and Hexagram

4.3.1 Construction and Symbolism

A hexagon inscribed in a circle has its side equals to the radius of the circle. This can be easy proved since the hexagon consists of 6 triangles packed together in the circle. The apex angle of each triangle is 60° (360°/6), and the triangle is isosceles, so the remaining angles must be 60°. Hence it is equilateral with the base equal to the other sides, i.e. the radius.

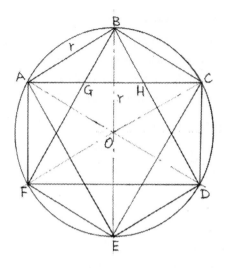

Figure 4.9 <u>Hexagon and Hexagram</u>

The hexagram can be constructed from the hexagon by connecting the diagonals except the 3 opposite pair of vertices. It can also be formed from 2 equilateral triangles, one pointed up and one down, interlaced with each other. The superimposition is such that 6 small equilateral triangles are formed at the corners.

The hexagram has long been used as a religious and community symbol, as well as fortune and magic. Known as the Star of David or in Hebrew "Magen David", it is an emblem representing Judaism and Jewish, and used by Israel on their national flag. Thought to come from the shape or emblem of King David's shield, there is so far no concrete evidence. In fact, the hexagram is a common symbol in the Middle East and North Africa, being a sign of good luck. It was also been used by ancient Hindus in their religious motifs, by magicians and alchemists in magic, witchcraft and sorcery, and astrologers in zodiacal horoscopes. Christian churches have displayed it as a star or festive decoration, especially during Christmas.

4.3.2 Geometric Properties

Referring to figure 4.9, by removing the 6 equilateral triangles from the corners of the hexagram, a reduced hexagon remains. A further reduced hexagram can be obtained in this way and so on. Hence like the pentagon and pentagram, the hexagon and hexagram are fractal.

For the diagonal AC, it is cut into 3 sections AG, GH and HC. From Δ ABC,

$$AC = 2r \cos 30° = 2r (\sqrt{3}/2)$$
$$= \sqrt{3} \; r$$

From   Δ AGB,
                AG = (r /2)/ cos30° = r/√3
Since   Δ AGB is isosceles, GB = AG
Also,   Δ BHG is equilateral, GH = GB, hence AG = GH
Similarly,   HC = GH
Therefore   AG = GH = HC
And AC is cut into 3 equal parts.

Check: r/√3 + r/√3 + r/√3 = √3 r

4.4   Some Related Findings

There have been many mathematical funs and findings related to the golden ratio and the golden geometry. Only 5 will be discussed below.

4.4.1   Constructing a Pentagon from the golden triangles

A pentagon can be constructed by combining a golden triangle with 2 golden gnomons, one on each side. In the following figure, ACD is the golden triangle, and ADE and ABC are the golden gnomons. The side of the pentagon is 1, being the base of ACD.

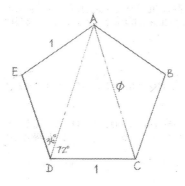

4.4.2   Construction of the Pentadecagon

A pentadecagon is a 15-sided polygon. This method is derived from Euclid's *Elements*, Book IV, Proposition 16.

An equilateral triangle and a pentagon are inscribed in a circle with a common vertex. The other vertices of both figures then divide the circle into 3/15, 2/15 and 1/15 along the circumference. Take the 1/15 (AB) as one side of the pentadecagon and bisection is not needed. Complete the other sides with a pair of compasses.

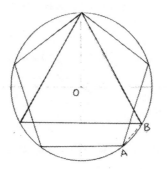

4.4.3 Platonic Solids

Euclid in the *Elements* proposed and proved that there exists only 5 regular solids, and each face of the solid is an identical regular polygon. These 5 solids are known as Platonic solids, after Plato.

(1) Tetrahedron   - 4 triangular faces
(2) Cube   - 6 square faces
(3) Octahedron   - 8 triangular faces
(4) Dodecahedron - 12 pentagonal faces
(5) Icosahedron   - 20 triangular faces

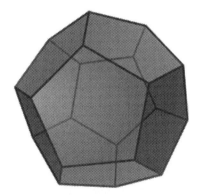

Figure 4.10 Dodecahedron
(Source: Pixabay)

4.4.4 Determine ϕ and $\sqrt{3}/2\phi$ from an Equilateral Triangle

This is part of the proposition of *The Elements* (Book XIII, 17) to inscribe a dodecahedron in a sphere. ϕ and $\sqrt{3}/2\phi$ are lengths required for the construction.

Inscribe an equilateral triangle ABC in a circle with centre O and radius = 1.

Bisect the base angles to obtain D and E, mid-point of AB and AC respectively. Extend ED to the circle, intercept at F. Draw a vertical line from F. Extend EB to this line, meet at G.

From figure 4.11, consider △ OAE

$$AE = OA \cos 30° = \sqrt{3}/2$$

$$EC = AE$$

$$BC = AC = 2AE$$

By similar triangles, $DE = \frac{1}{2}BC = AE$
$$= \sqrt{3}/2$$

Let DF = x,

By symmetry,   EH = DF

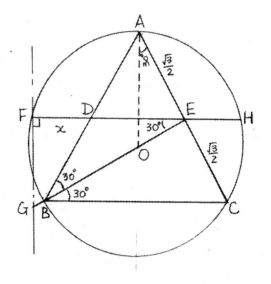

Figure 4.11 <u>Determine ϕ and $\sqrt{3}/2\phi$ from an Equilateral Triangle</u>

Chord FH intersects with chord AC,

Hence $(FE)(EH) = (AE)(EC)$

$$(x + \sqrt{3}/2)x = (\sqrt{3}/2)(\sqrt{3}/2) = 3/4$$

$$x^2 + \sqrt{3}/2x - 3/4 = 0$$

$$x = \frac{-\sqrt{3}/2 \pm \sqrt{\frac{3}{4} + 4\left(\frac{3}{4}\right)}}{2} = \sqrt{3}/2 \left(\frac{\sqrt{5}-1}{2}\right)$$

(neglect the negative root)

$$= \sqrt{3}/2 \; \phi_2$$

And $EF = x + DE = \sqrt{3}/2 \; \phi_2 + \sqrt{3}/2$

Hence $EF = \sqrt{3}/2 \; (\phi_2 + 1) = \underline{\sqrt{3}/2 \; \phi}$

And $EG = EF / \cos 30° = \underline{\phi}$

4.4.5 Cone formed from the 3,4,5 Triangle

Construct a hollow cone with apex O, base radius (a) = 3 and height (h) = 4. Then the length of the inclined surface (r) = 5, being the 3,4,5 triangle.

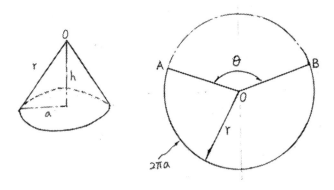

Now cut the cone along (any) r and open it, forming a fan shape with O at the tip or root. This is called the development (on a sheet of paper / metal) of the hollow shell. The fan has a radius of r, and arc length 2πa, being the circumference of the cone base. The angle α of the fan is given by arc length divided by radius.

$$\alpha = \frac{2\pi a}{r}$$

α cannot be 360° (2π), because then r = a, and the cone degenerates to a flat circle. The remaining angle is

θ = 2π - α
　= 2π - 2π(3)/5 = 2π (2/5)
　= 360° (2/5)
　= 144°　　　double angle of golden triangle

If　　a = 4, h = 3 (r = 5)
Then　θ = 2π - 2π(4)/5 = 360° (1/5)
　　　 = 72°　　　angle of golden triangle!

Chapter 5

Vesica Piscis and Spirals

This chapter may be treated as a continuation of the last one.

5.1   Vesica Piscis

5.1.1   What is Vesica Piscis

The Vesica Piscis (Pisces) is the area overlapped by 2 circles of the same radius with the circumference touching the centre of each other. In Latin the meaning is "fish bladder". It has been regarded as a religious divine symbol in Europe and Asia, also known as the *Mandorla*. In medieval church, the Vesica Piscis is a standard motif, dedicated to the Virgin Mary. It also represents aureola - radiance of luminous light, enclosing Jesus or Virgin Mary. The same purposes have been used for Buddha, and some olden time Persian kings. In other religions or beliefs, Vesica Piscis is a symbol for the offspring from god and goddess. An interesting recent finding is that photographs from the Hubble Space Telescope show that the Hour-glass Nebula is in the form of a Vesica Piscis.

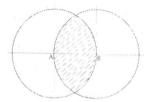

5.1.2 Geometry of Vesica Piscis

A rhombus ACBD is constructed joining the 2 intersections C, D and centres A, B of the 2 circles forming the Vesica Piscis. Note that the width of this rhombus AB is the radius of the circle and let it be 1 unit.

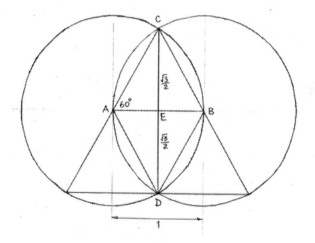

Since the sides of the rhombus are equal to the radius, so it is in fact consists of two equilateral triangles (with all angles 60°). Hence the angles of the rhombus are 60° and 120°, and its height CD is $\sqrt{3}$.

The Vesica Piscis and its associated rhombus also exhibit fractal nature. To produce the reduced figures, bisect the half 120° (60°) of the rhombus on both sides to obtain the vertices of the new rhombus. The new Vesica Piscis is formed by using the width of the new rhombus as the radius to complete the arcs, with centres at the vertices.

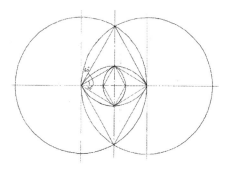

Figure 5.1 <u>Fractal nature of Vesica Piscis</u>

5.1.3 √2, √5 and the Pentagon

Besides the equilateral triangle, the 1:1 and 1:2 triangles can be formed inside the Vesica Piscis. Refer to figure 5.2, Δ BAD is 1:1, hence BD= √2. Δ OAD is 1:2 and OD= ½√5, QD=√5. Therefore, √2, √3 and √5 are all contained inside the Vesica Piscis, where many ancient proportions were derived. Also, a pentagon can be constructed from it. Take O as centre, swing OD to the horizon at H.

$$AH = OH - OA$$
$$= \tfrac{1}{2}(\sqrt{5} - 1)$$

If we form an isosceles Δ AEH with base AH and sides HE=AE=1, then AEH is a golden triangle. Bisect AH to get E, AE = radius of circle = 1. Hence ∠EAH = 72° and ∠BAE = 108°, interior angle of pentagon. Use AE or BA, which is the radius of circle, to complete the figure BAEGF. It is a regular pentagon of side equals to radius of the Vesica Piscis.

By drawing a third circle of the same radius with centre at P (lowest point of the Vesica Piscis), intersections with one circle and the vertical line are obtained at M and N. Joining M and N and extend upward to the other circle, point E can be determined geometrically. This is a quicker step.

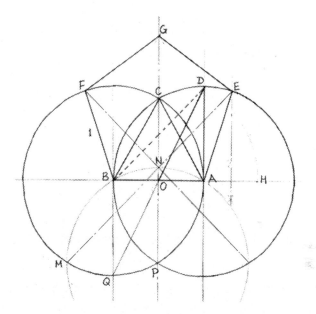

Figure 5.2 <u>Constructing a Pentagon</u>
<u>from the Vesica Piscis</u>

5.1.4   The 4-Point Star

Let the radius be 1 unit, draw a circle with centre O. Add 4 outer circles of the same radius, with centres at right angle touching the central circle. Take one such centre D, draw a construction (faint) line from D to E and extend to A, intersecting the vertical centre line. Repeat with the other 3 centres. Complete the 4-point star as shown in figure 5.3.

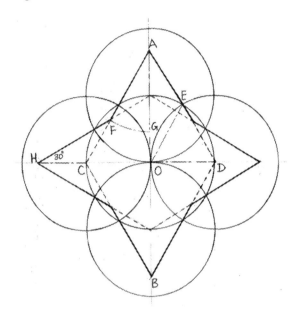

Figure 5.3 <u>The 4-Point Star from the Vescia Piscis</u>

This 4-point star is in fact formed by 2 intersecting rhombuses of the Vesica Piscis at right angle to each other. The base of the rhombus is 2, which is radius of the circle forming the larger Vesica Piscis (not shown). To prove this, since OE is the radius of O and DO and DE radii of D, Δ ODE is equilateral. $\angle$ODE is 60°.

$$OA = OD \tan60° = \sqrt{3}$$

$$AB = 2\ OA = 2\sqrt{3}$$

Also, $CD = 2\ OD = 2$

Therefore rhombus ADBC is one formed from a Vesica Piscis of radius 2.

Let us look at the line AC, cut at point F.

AC = CD, sides of the equilateral Δ ACD.

Therefore   AC = 2

Since   ∠ACB = 120°,

$\angle FCH = \frac{1}{2}(360° - 120°) = 120°$

∠CHF = 30°   (by symmetry),

In Δ CFH, ∠HFC = 30° (180° – 120° – 30°)

Therefore   Δ CFH is isosceles

And   FC = CH

Now   CH = OH – OC = √3 – 1

Hence   FC = √3 - 1

AF = AC – FC

= 2 – (√3 – 1)

= 3 – √3

Analogy to the golden ratio, AC is the whole line, AF is the greater sector and FC the smaller sector. Let us examine the ratios.

AC : AF = 2 : (3 – √3)

AF : FC = (3 – √3) : (√3 – 1)

$$\frac{AC}{AF} = \frac{2}{3 - \sqrt{3}} = \frac{2\left(3 + \sqrt{3}\right)}{\left(3 - \sqrt{3}\right)\left(3 + \sqrt{3}\right)} = \frac{2\left(3 + \sqrt{3}\right)}{6}$$

$$= 1 + 1/\sqrt{3}$$

$$\frac{AF}{FC} = \frac{3-\sqrt{3}}{\sqrt{3}-1} = \frac{\left(3-\sqrt{3}\right)\left(\sqrt{3}+1\right)}{\left(\sqrt{3}-1\right)\left(\sqrt{3}+1\right)} = \frac{3-3+3\sqrt{3}-\sqrt{3}}{2}$$

$$= \sqrt{3}$$

There is not much significance. However, let us swing AF to the vertical line to get AG.

Now $AO = \sqrt{3}$, $AG = AF = 3 - \sqrt{3}$

$$GO = AO - AG$$
$$= \sqrt{3} - (3 - \sqrt{3})$$
$$= 2\sqrt{3} - 3$$

Then $\dfrac{AO}{AG} = \dfrac{\sqrt{3}}{3-\sqrt{3}} = \dfrac{\sqrt{3}\left(3+\sqrt{3}\right)}{9-3}$

$$= \frac{\sqrt{3}+1}{2}$$

$\dfrac{AG}{GO} = \dfrac{3-\sqrt{3}}{2\sqrt{3}-3} = \dfrac{\left(3-\sqrt{3}\right)\left(2\sqrt{3}+3\right)}{3}$

$$= \sqrt{3}+1$$

And   AG : GO = 2 (AO : AG)

So there is an interesting ratio in cutting a line:

The greater sector to the lesser equals to twice the whole to the greater sector.

Put it in an algebraic equation,

Let the greater sector = x,   the lesser sector = 1

Then $\dfrac{x}{1} = 2\,\dfrac{1+x}{x}$

$x^2 = 2 + 2x$

$x^2 - 2x - 2 = 0$

And $x = \dfrac{2 \pm \sqrt{4+8}}{2}$

or $x_1 = \sqrt{3} + 1$ neglecting the negative root

This matches the former calculation.

If the whole line (AO) = 1,   greater sector = x

Hence   lesser sector = 1 − x,   x < 1

Then $\dfrac{x}{1-x} = 2\,\dfrac{1}{x}$

$x^2 = 2 - 2x$

$x^2 + 2x - 2 = 0$

And $x = \dfrac{-2 \pm \sqrt{4+8}}{2}$

or $x_2 = \sqrt{3} - 1$ neglecting the negative root

Also, $1/(\sqrt{3} - 1) = \frac{1}{2}(\sqrt{3} + 1)$

or $\dfrac{1}{x_1} = \frac{1}{2}\,x_2$

5.2   Logarithmic Spiral

5.2.1   General Form

A spiral is a 2 dimensional curve that coils from a point of a co-ordinate system and expands outward or contracts inward. The polar (i.e. in polar co-ordinates r and θ) equation of a logarithmic spiral is given by

$$r = ae^{b\theta}$$

where   r = distance from origin

θ = angle from x-axis

a, b = constants

Note that when θ = 0, r = a. Hence the spiral cannot start from 0 or the origin. This may mean that growth cannot start from nothing, initially there has to be a dimension. At successive points along the spiral, the radii (distance from the origin) are in geometric progression.

Logarithmic spiral is also known as growth spiral, equiangular spiral and spira mirabilis (named by Jacob Bernouli, meaning marvellous spiral). The term equiangular was introduced by Rene Descartes, who is generally accepted as the first person to describe the spiral. If a line is drawn from the origin (pole) through any point on the spiral, it cuts the curve at the same angle with the tangent at that point. This property can be used to construct the spiral (refer to figure 5.5).

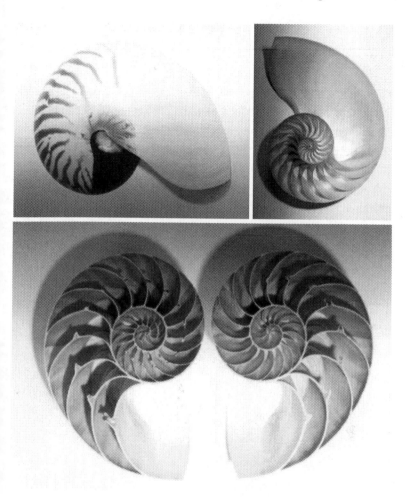

Figure 5.4 <u>A Nautilus Shell and Bisected Halves</u>

The size of the spiral increases while its shape unaltered, a property known as *self-similarity*. This phenomenon appears in nature, as in the nautilus and mollusks shells. Other forms include arms of spiral galaxies, bands of tropical cyclones and nerves of the human cornea. Horns of many animals also show this property, but in this case it is like a spiral cone. The tip rather than the open end keeps on curling inward.

Figure 5.4 shows a nautilus shell which is subsequently bisected neatly into 2 halves. The interior shape is a logarithmic or growth spiral.

The logarithmic spiral can be constructed by bisecting a golden triangle. Details have been covered in Chapter 4, Section 4.1.3. Another method is to construct equally spaced rays (Figure 5.5). Start at one point along the ray and draw perpendicular line to meet the next ray. Then from this intersection draw another perpendicular and so on. As number of rays approaches infinity, the curve approaches a smooth spiral.

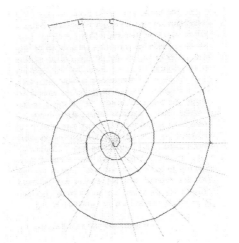

Figure 5.5 <u>Logarithmic Spiral Constructed by Drawing Perpendiculars to Rays</u>

One approximate method is to start with the 1:1 triangle.

Then continue to draw right angle triangles with 1 as the opposite side and the hypotenuse of the previous one as the base (figure 5.6). The sides of the triangles approximate to the spiral.

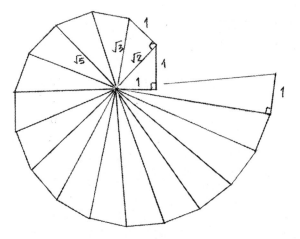

Figure 5.6 <u>Approximate Logarithmic Spiral</u>
<u>from Right Angle Triangles</u>

5.2.2 The Golden Spiral

The golden spiral is a logarithmic spiral that grows outward by the golden ratio for every 90° ($\pi/2$) of rotation.

From general equation

Consider 2 radii r_1, r_2 at 90° apart.

$$r_1 = ae^{b\theta}$$

$$r_2 = r_1\phi = ae^{b\theta_2}$$

$$\frac{r_2}{r_1} = \phi = e^{b(\theta_2 - \theta_1)} = e^{b\pi/2}$$

$$\ln \phi = \pi b/2$$

$$b = 2 \ln\phi /\pi$$

$$= 0.3063$$

Let a = 1

Equation of the golden spiral becomes

$$r = e^{0.3063\theta}$$

The plotted graph is shown in Figure 5.7.

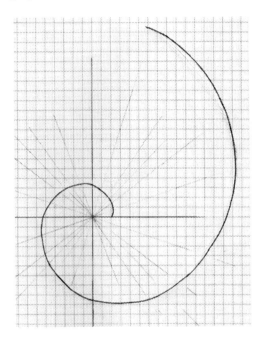

Figure 5.7 <u>Plotted Golden Spiral</u>

The golden spiral can also be approximated from the golden rectangle. The curve is very similar to the Fibonacci spiral, as the golden rectangle and the Fibonacci numbers are closely related.

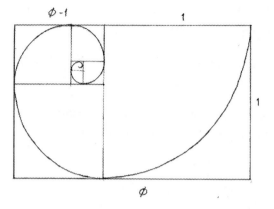

Figure 5.8 <u>Approximate Golden Spiral from Golden Rectangles</u>

5.3 Arithmetic Spiral

It is the locus of a point moving outward (or inward) from the origin or fixed point, with a constant speed while at the same time rotating at constant angular speed. It is also known as Archimedean spiral or Spiral of Archimedes. The coils of balance springs in watch and the grooves of early gramophone records are arithmetic spirals.

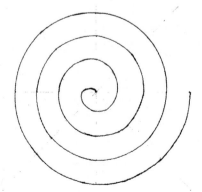

The simplified general formula is

$$r = a + b\theta$$

where a, b are constants.

Similar to the golden spiral, let us try to construct an arithmetic spiral which increases by ϕ for every 90° turn.

Then $\quad r_1 = a + b\theta$

$$r_2 = (1 + \phi)r_1 = a + b(\theta + \pi/2)$$

$$r_2 - r_1, \quad r_1\phi = b\pi/2$$

Initially $\theta = 0, r = a$

Hence $\quad a\phi = b\pi/2$

Or $\quad b / a = 2\phi/\pi$

Let $\quad a = 1$

$$b = 2\phi/\pi = 1.03$$

The resulting equation is

$$r = 1 + 1.03\theta$$

Figure 5.9 shows the plotted graph. Compare with the golden spiral (figure 5.7), the initial growth is faster but is gradually overtaken by the golden spiral.

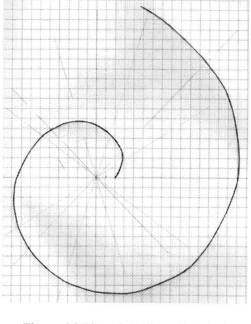

Figure 5.9 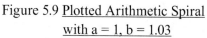 Plotted Arithmetic Spiral
with a = 1, b = 1.03

Chapter 6

Fibonacci Numbers and Continued Fraction

6.1   Fibonacci Numbers

6.1.1   Fibonacci Sequence

In the early 13ᵗʰ century, year 1202, an Italian mathematician Fibonacci, also known as Leonado of Pisa, published the *liber Abai,* which means "Book of Abacus". The book is actually about arithmetic and algebra, and includes the following problem:

"How many pairs of rabbits will be produced in a year beginning with a single pair, if in every month each pair produces one new pair. The parent and new pairs begin to give offspring after one month."

Let n be the nᵗʰ month and N be the number of rabbit pairs.

Start with one pair in the first month.

$n = 1$ and $n = 2$, $N = 1$

$n = 3$, the parent pair produces a new pair, so $N = 2$.

$n = 4$, the parents continue to produce another pair, $N = 3$.

$n = 5$, the parents and the 1st offspring produce a new pair each,
      $N = 5$.

$n = 6$, parents, 1st and 2nd offspring produce a new pair each,
      $N = 8$ ......

Ignore about genetics and mortality, the outcome sequence

$$1, 1, 2, 3, 5, 8, 13 \ldots\ldots$$

is what is known as Fibonacci numbers. Each number in the sequence is the sum of the 2 preceding numbers, starting with 1, 1. (So the solution to Fibonacci's original problem is 143 pairs - 144 minus the parent pair.)

Denote $= n^{th}$ Fibonacci number,

Thus $F_1 = 1$, $F_2 = 1$,

$$F_3 = F_2 + F_1 = 2$$
$$F_4 = F_3 + F_2 = 3$$
$$F_5 = F_4 + F_3 = 5$$
$$F_6 = F_5 + F_4 = 8$$

$$\vdots \qquad \vdots \qquad \vdots$$

$$\vdots \qquad \vdots \qquad \vdots$$

$$F_n = F_{(n-1)} + F_{(n-2)}$$
$$F_{(n+2)} = F_{(n+1)} + F_n$$

n = natural number

This is a form of linear recurrence equation, and without going through the complicate theories and procedures, the solution is given by

$$F_n = \frac{1}{\sqrt{5}}\left[\left(\frac{1+\sqrt{5}}{2}\right)^n - \left(\frac{1-\sqrt{5}}{2}\right)^n\right]$$

This is known as Binet's formula.

Note that $\left(\frac{1+\sqrt{5}}{2}\right)^n$ is ϕ^n and $\left(\frac{1-\sqrt{5}}{2}\right)^n$ is $(-\phi)^n$

Hence the Fibonacci numbers are closely related to the golden ratio.

Two similar sequences are introduced below.

(1)   Lucas numbers, L_n

Here $L_1 = 1$, $L_2 = 3$, and
$$L_n = L_{n-1} + L_{n-2}$$

The sequence is   1, 3,4, 7, 11, 18, 29 ……..

Note that two subsequent odd or even numbers added together are in multiple of 5 or 10.

(2)   Pell numbers, P_n

This time $P_0 = 0$, $P_1 = 1$, and
$$P_n = 2P_{n-1} + P_{n-2}$$

The sequence is   1, 2, 5, 12, 29, 70, 169 …….

6.1.2   Some Identities

Many mathematical funs and identities can be derived from the Fibonacci numbers. Just a few are listed below.

(1) $\qquad 2F_n F_{n+1} = F_{n+1}^2 + F_n^2 - F_{n-1}^2$

Proof

From $\quad F_{n+1} = F_n + F_{n-1}$

$\qquad \left(F_{n+1} - F_{n-1} \right)^2 = F_{n-1}^2$

$\qquad F_{n+1}^2 - 2F_{n+1} F_n + F_n^2 = F_{n-1}^2$

$\qquad 2F_n F_{n+1} = F_{n+1}^2 + F_n^2 - F_{n-1}^2$

Hence $\quad 2F_n F_{n+1} = F_{n+1}^2 + F_n^2 - F_{n-1}^2$

(2) $\qquad F_1 + F_2 + F_3 + \ldots + F_n = F_{n+2} - 1$

The *Method of Induction* for proving a theorem / proposition with n terms is introduced here.

First test that the proposition is true for n = 1. If so, let it be true for n (formally m should be used instead of n) terms. Then test that it is true for n+1 (m+1) terms. Once this step is positive, the proposition is valid.

For n = 1, $\qquad F_1 = 1 = F_3 - 1$

(n = 2, 3 $\qquad F_1 + F_2 = 2 = F_4 - 1$

$\qquad\qquad F_1 + F_2 + F_3 = 4 = F_5 - 1$)

Let $F_1 + F_2 + F_3 + \ldots + F_n = F_{n+2} - 1$ be true

$F_1 + F_2 + F_3 + \ldots + F_n + F_{n+1}$

$\qquad = \left(F_{n+2} - 1 \right) + F_{n+1}$

$\qquad = \left(F_{n+2} + F_{n+1} \right) - 1$

$\qquad = F_{n+3} - 1$

(3) $\qquad F_1^2 + F_2^2 + F_3^2 + \ldots + F_n^2 = F_n F_{n+1}$

Proof

For n = 1 and n = 2,

$$F_1^2 = 1 = F_1 F_2$$

$$F_1^2 + F_2^2 = 2 = F_2 F_3$$

Assume it is true for n terms. Consider n+1 terms:

$$F_1^2 + F_2^2 + F_3^2 + \ldots + F_n^2 + F_{n+1}^2$$

$$= F_n F_{n+1} + F_{n+1}^2 = F_{n+1}\left(F_{n+1} + F_n\right)$$

$$= F_{n+1} F_{n+2}$$

(4) $\qquad F_n^2 + F_{n+1}^2 = F_{2n+1}$

Proof

For n = 1, and 2

$$F_1^2 + F_2^2 = 2 = F_3$$

$$F_2^2 + F_3^2 = 5 = F_5$$

Assume it is true for nth terms.

i.e. $F_n^2 + F_{n+1}^2 = F_{2n+1}$

Consider the n+1th term:

$$F_{n+1}^2 + F_{n+2}^2 = F_{n+1}^2 + \left(F_{n+1} + F_n\right)^2$$

$$= F_{n+1}^2 + F_{n+1}^2 + F_n^2 + 2F_{n+1}F_n$$

From (1) above,

$$2F_n F_{n+1} = F_{n+1}^2 + F_n^2 - F_{n-1}^2, \text{ hence}$$

$$F_{n+1}^2 + F_{n+2}^2 = 3(F_{n+1}^2 + F_n^2) - (F_n^2 + F_{n-1}^2)$$

$$= 3F_{2n+1} - F_{2n-1}$$

$$= F_{2n+1} + F_{2n+1} + (F_{2n+1} - F_{2n-1})$$

$$= F_{2n+1} + (F_{2n+1} + F_{2n})$$

$$= F_{2n+1} + F_{2n+2}$$

$$= F_{2n+3}$$

The following 2 identities will not be proved.

(5) $\quad F_n = \dfrac{1}{5}(L_{n-1} + L_{n+1})$

(6) $\quad F_{n+1} = \dfrac{1}{2}(F_n + L_n)$

6.2 Fibonacci Numbers and the Golden Ratio

6.2.1 Relation between Fibonacci Numbers and ϕ

For the Fibonacci sequence 1, 1, 2, 3, 5, 8, 13, 21, 34, 55 ….,

Try to evaluate $\quad \dfrac{F_{n+1}}{F_n}$

Start with n = 5, $\quad \dfrac{8}{5} = 1.6$

$$\dfrac{13}{8} = 1.625$$

$$\frac{21}{13} = 1.615335$$

$$\frac{34}{21} = 1.619$$

$$\frac{55}{34} = 1.61765$$

It can be seen that as n approaches infinity (n→∞),

$$\frac{F_{n+1}}{F_n} \to \phi$$

For the Lucas sequence 1, 3, 4, 7, 11, 18, 29, 47 ….

$$\frac{L_5}{L_4} = \frac{11}{7} = 1.5714$$

$$\frac{L_6}{L_5} = \frac{18}{11} = 1.6364$$

$$\frac{L_7}{L_6} = \frac{29}{18} = 1.6111$$

$$\frac{L_8}{L_7} = \frac{47}{29} = 1.6206$$

Hence   as n →∞, $\dfrac{L_{n+1}}{L_n} \to \phi$

We are going to examine for any linear recurrence sequence

$$S_n = S_{n-1} + S_{n-2},$$

The ratio of $\dfrac{S_{n+1}}{S_n}$ as n→∞

Let us do it formally. Put x $= \dfrac{S_{n+1}}{S_n}$

If x converges to a fixed value, then it should be bounded by a limit, i.e. $x \le M$, where M is a constant.

Since $\dfrac{S_{n+1}}{S_n} = \dfrac{S_n + S_{n-1}}{S_n}$

For all initial conditions S_1, S_2 ...., so far the net summation is positive, $S_{n-1} \le S_n$

Therefore $\quad x = \dfrac{S_n + S_{n-1}}{S_n} \le \dfrac{S_n + S_n}{S_n} = 2$

So x is bounded and we can evaluate it.

As n→∞, $\quad x = \dfrac{S_{n+1}}{S_n} = \dfrac{S_n}{S_{n-1}}$

And $\quad x = \dfrac{S_{n+1}}{S_n} = \dfrac{S_n + S_{n-1}}{S_n}$

$$= 1 + \dfrac{S_{n-1}}{S_n} = 1 + \dfrac{1}{x}$$

$$x^2 = x + 1$$

Hence $\quad x^2 - x - 1 = 0$

This is the familiar equation, and the solution is of course

$$x = \phi$$

6.2.2   Power of ϕ

Recall $\quad \phi^2 = \phi + 1 = F_2\, \phi + F_1$

$$\phi^3 = \phi(\phi + 1) = \phi^2 + \phi$$
$$= (\phi + 1) + \phi$$
$$= 2\phi + 1 = F_3 \phi + F_2$$
$$\phi^4 = \phi(2\phi + 1) = 2\phi^2 + 1$$
$$= 3\phi + 2 = F_4 \phi + F_3$$

It seems that $\qquad \phi^n = F_n \phi + F_{n-1}$ ....(7)

Proof by induction

Let ϕ^n be true.

Consider $\phi^{n+1} = \phi \, \phi^n$
$$= \phi(F_n \phi + F_{n-1})$$
$$= F_n \phi^2 + F_{n-1} \phi$$
$$= F_n (\phi + 1) + F_{n-1} \phi$$
$$= (F_n + F_{n-1})\phi + F_n$$
$$= F_{n+1} \phi + F_n$$

Therefore equation (7) is true for all values of n.

Note that ϕ^2 can also be written as a recurrence sequence.

Since $\qquad \phi^2 = \phi + 1$
$$\phi^3 = \phi(\phi + 1) = \phi^2 + \phi$$
In general $\quad \phi^n = \phi^{n-1} + \phi^{n-2}$

----

which is in the form $\qquad S_n + S_{n-1} + S_{n-2}$

6.2.3   Fibonacci Numbers and the Pythagorean Triples

The last part of Table 3.1 Pythagorean Triples is extracted below. Note that when the generating seeds m and n are in consecutive Fibonacci numbers and as m and n increase, a and b of the Pythagorean triples are close to the ratio of 1 : 2. That is the resulting right angle triangle is 1 : 2, which is closely related to the golden ratio ϕ.

m	n	a	b	c
8	5	39	80	89
13	8	105	208	233
34	21	715	1428	1597

6.2.4   Fibonacci Numbers and the Golden Geometry

Refer to the golden triangle ABC below,

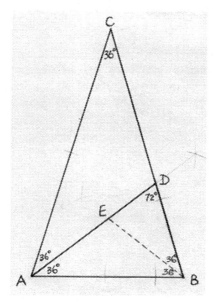

Figure 6.1 <u>Fibonacci 5-line series</u>

The golden triangle is bisected at one of the base angle in fractal shapes twice, and

CB = CD + DB

Now   CB is one inclined side of Δ ABC

CD = AD (isosceles Δ ACD), and

AD is same inclined side of Δ ABD

DB is same side of Δ BDE

Thus   CB = CD + DB

is same as

$$S_n + S_{n-1} + S_{n-2}$$

And figure 6.1 is known as "Fibonacci 5-line series".

6.2.5   Fibonacci Spiral

The Fibonacci spiral is constructed from a rectangle formed by two consecutive Fibonacci numbers. First divide the rectangle into a square and a smaller rectangle. Then subdivide the latter into yet a square and a remaining rectangle, and so on, until it reaches the 1x1 square. Note that all the squares have side in Fibonacci numbers.

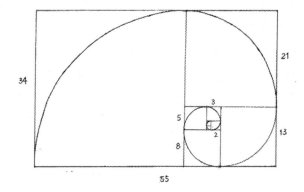

Figure 6.2 Fibonacci Spiral

Use the corner of the square as centre and side as radius, draw a 90° arc across each square as shown in figure 6.2. The resulting curve is a spiral.

Note that the Fibonacci spiral is almost same as the spiral constructed from the golden rectangle (refer to figure 5.8, which is just upside down). Hence the Fibonacci spiral is an approximation of the golden spiral.

6.3   Plant Growth and Golden Angle

6.3.1   The Golden Angle

In 1754, Swiss botanist Charles Bonnet first explained the mechanism of leaf growth - order of leaf placement, known as phyllotaxis. The leaves are spaced around the stem in an upward spiral. Each leaf grows above one another by more or less a constant angle, known as the *divergence angle*, i.e. the leaves are angularly equally placed along the generative spiral. In most species, this angle is about 137.5°. This is called *golden angle*.

Why 137.5° ? Most text would reveal that

137.5° = 360°(1 - ϕ_2), but there should be more than that.

Recall Euclid's mean and extreme proportion

$$\frac{x}{1} = \frac{1-x}{x} \qquad \text{where x < 1}$$

Let the line to be cut is a unit and y is the lesser segment

Then
$$\frac{a-y}{a} = \frac{y}{a-y}$$

$$(a - y)^2 = ay$$

$$a^2 - 3ay + y^2 = 0$$

$$y = \frac{3a \pm \sqrt{9a^2 - 4a^2}}{2}$$

$$= a\left(\frac{3 \pm \sqrt{5}}{2}\right)$$

The positive square root is neglected as y will be > a

Hence $y = a\left[1 - \left(\frac{\sqrt{5}-1}{2}\right)\right]$

$$= a (1 - \phi_2) \qquad \text{as expected}$$

Take $a = 2\pi$ (360°), circumference of a circle,

then $y = 360° (1 - \phi_2) = 137.508°$

$$\cong 137.5°, \qquad \text{the golden angle}$$

Thus the golden angle divides the circumference of a circle in the golden ratio.

In fact, the same result can be obtained by multiplying ϕ_2 by 360° to get 222.492°, and subtract it from 360°. In geometry, if an angle $\theta > 180°$, we usually consider the angle 360° - θ.

6.3.2   Primordia

At the tip of the shoot of a growing plant is a region called meristem. Here cells are multiplied rapidly. Just below the advancing tip, the apex, tiny side buds called primordia begin to protrude one by one. The apex continues to grow upward away from each primordium. Eventually the buds develop into leaf, petal, sepal or floret.

The locations of the primordia trace out a spiral when seen from above. Usually 2 spirals can be spotted, one clockwise and one counterclockwise. The double spiral pattern is more easily caught when the primordia develop into florets in a flower head, since in this case they appear almost on the same plane. These florets are just modified leaves. Most common, the heads of sunflowers, daisies and cauliflowers show this pattern. What is more interesting, the quantities of the 2 twisting spirals are in consecutive Fibonacci numbers. For example, the sunflowers have spirals of 55/34, 89/55 and 144/89.

Figure 6.3 <u>Floret of Sunflower Showing the Spirals</u>
<u>(Source: Pixabay)</u>

Not only florets exhibit Fibonacci double spiral, scales of fruits such as pineapples and pine cones also appear in spirals. The hexagonal scales on their surfaces are packed in spiral fashion. A pine cone is shown in figure 6.4, (a) the cone before "burst" out, (b) after bursting the spirals can still be spotted, and (d) top view after bursting. The divergence angle between consecutive "petals" is about 72°.

Figure 6.5 is the bottom view of a burst cone (larger one), with the scales still packed in double spiral. Two spirals are marked, "x" is clockwise and "Δ" is counterclockwise. Are they in Fibonacci numbers?

(a)   (b)   (c)

(d)

Figure 6.4 <u>Different Views of Pine Cone</u>

Figure 6.5 <u>Bottom View of a "Burst" Pine Cone</u>

6.3.3   Compact Packing

In 1907, Dutch botanist Gerrit van Iterson tested the divergence angle by plotting successive points separated by 137.5° on a tightly wounded spiral. Because of the effect of misalignment of neighbouring points, the human eyes pick out two families of intertwisting spirals, one winding clockwise and the other counterclockwise. He explained that as the Fibonacci numbers and golden section are related, the numbers of spirals are consecutive Fibonacci numbers.

German scientist Helmut Vogel performed numerical simulations in 1979, which showed that if successive primordia are placed along the generative spiral using the golden angle, they will be packed together most efficiently. To pack a spiral compactly, the divergence angle cannot be an exact factor of 360°, otherwise radial lines will form and there will be gaps. So the angle has to be an irrational multiple of 360°, or cannot divide it in exact value. ϕ is considered as the worst irrational number and would give maximum packing. Other experiments (Stephane Douady and Yves Couder, 1996) suggested that the observed phyllotaxis represents a state of minimal energy for a system of mutually repelling buds.

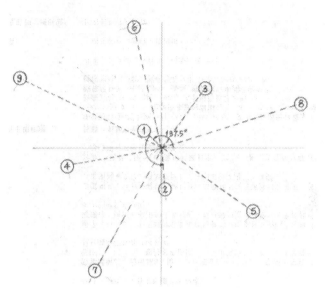

Figure 6.6 <u>Primordia separated by the Golden Angle</u>

One day the author carried out a small packing game using toothpicks. The diameter of the toothpick was roughly 2 mm and length 65 mm. Opening of the container had an average diameter of 26.6 mm with a clearance depth of 41 mm, slightly concave at the bottom. To leave minimum gaps, the toothpicks were slightly slanted, so that their cross-section became elliptic. Guess how many were packed?

Totally 144 pieces were tightly packed in the container, a Fibonacci number, or sum of two consecutive numbers 89/55. The angle of inclination of the toothpicks was about 16° - 18°, or 72° - 74° from the horizontal. Figure 6.7 shows the outlook of the results. Can the reader spot out the spirals?

Figure 6.7 <u>Compact Packing Experiment</u>

6.4 Continued Fraction

6.4.1 <u>Simple Continued Fraction</u>

A continued fraction is an expression of the form

$$\eta = a_0 + \cfrac{b_0}{a_1 + \cfrac{b_1}{a_2 + \cfrac{b_2}{a_3 + \cfrac{b_3}{a_4 + \cfrac{b_4}{a_5 + \dots}}}}}$$

$a_0 \ldots a_n, b_0 \ldots b_n$ may be real or imaginary numbers, and n may be finite or infinite. If, $b_1 = b_2 = \ldots = b_n = 1$ then η is a *simple continued fraction*.

Simple continued fraction may be written as

$$\eta = a_0 + \cfrac{1}{a_1 +} \cfrac{1}{a_2 +} \cdots \cfrac{1}{a_n}$$

or simply $\eta = \left[a_0, a_1, a_2, \ldots a_n \right]$

a_0 may be zero. If n terminates, i.e. finite, η is rational. If n goes to infinity, η is irrational. The best approximation to the irrational number can be obtained by truncating (cut down) the continued fraction at the suitable stage.

Continued fraction can be used to replace a quadratic equation or vice versa. For example

$$x^2 - 5x - 2 = 0$$

$$x = 5 + \frac{2}{x}$$

Substitute x to the RHS of the equality

$$x = 5 + \cfrac{2}{5 + \cfrac{2}{x}}$$

Repeating, $x = 5 + \cfrac{2}{5 + \cfrac{2}{5 + \cfrac{2}{x}}}$

 :

 :

Here, the fraction is infinite.

6.4.2 Simple Continued Fraction with all $a_i = n$

Then

$$\eta = n + \cfrac{1}{n + \cfrac{1}{n + \cfrac{1}{n + \cfrac{1}{n + \dots}}}}$$

We can treat this fraction in another way.

Form a sequence c_n in terms of n, starting with $c_0 = n$. The next value is n + reciprocal of the previous one. i.e.

$$c_0 = n$$

$$c_1 = n + \frac{1}{n}$$

$$c_2 = n + \cfrac{1}{n + \cfrac{1}{n}}$$

$$\vdots$$

$$c_{i+1} = n + \frac{1}{c_i}$$

The sequence will converge for $n \geq 0$, so as $i \to \infty$, $c_{i+1} = c_i$

Let $x = c_{i+1} = c_i$

Hence $x = n + \dfrac{1}{x}$

$$x^2 = nx + 1$$

$$x^2 - nx - 1 = 0 \dots\dots\dots\dots\dots\dots(8)$$

Ignore a_0 ($a_0 = 0$) and consider the fraction part only.

Form a new sequence d_n, with $d_0 = \dfrac{1}{n}$ and

d_{i+1} = reciprocal of $(n + d_i)$

Then $\quad d_0 = \dfrac{1}{n}$

$$d_1 = \cfrac{1}{n + \cfrac{1}{n}}$$

$$\vdots$$

$$d_{i+1} = \cfrac{1}{n + d_i}$$

Also, this sequence will converge for $n \geq 0$.

Let $\qquad\qquad x = d_{i+1} = d_i$

Hence $\qquad\qquad x = \dfrac{1}{n + x}$

$$x^2 + nx = 1$$

$$x^2 + nx - 1 = 0 \quad\ldots\ldots\ldots\ldots\ldots\ldots\ldots(9)$$

Let $n = 1$, equation (8) becomes

$$x^2 - x - 1 = 0$$

and the solution is $\qquad x = \phi$

Therefore $\qquad \phi = 1 + \cfrac{1}{1 + \cfrac{1}{1 + \cfrac{1}{1 + \ldots}}}$

And equation (9) becomes

$$x^2 + x - 1 = 0$$

The solution is $x = \phi_2$

Thus
$$\phi_2 = \cfrac{1}{1 + \cfrac{1}{1 + \cfrac{1}{1 + \ldots}}}$$

6.4.3   Rational Numbers

A number is said to be rational if it can be expressed as a fraction p/q, where p, q are integers.

A rational number can be written as continued fraction by repeated division in the following steps:

(1) If p ≥ q, then $a_0 \geq 1$, otherwise = 0.

Divide p by q, the quotient is a_0, and remainder r_0.

(2) Divide q by r_0 or if p < q, q by p. Quotient is a_1 and remainder r_1.

(3) Divide r_0 or p by r_1, quotient is a_2 and remainder r_2.

(4) Divide r_1 by r_2 to get a_3 and r_3.

(5) Continue the division until the remainder equals to zero.

Note that any finite decimal figures can be written as a fraction. For example

$$3.1416 = 3 + \frac{1416}{10000}$$

Working examples:

1) $\dfrac{25}{11} = 2 + \dfrac{3}{11}$ $a_0 = 2$

$\dfrac{11}{3} = 3 + \dfrac{2}{3}$ $a_1 = 3$

$\dfrac{3}{2} = 1 + \dfrac{1}{2}$ $a_2 = 1$, and $a_3 = 2$

Therefore $\dfrac{25}{11} = [2,3,1,2]$

It can be seen that $\dfrac{1}{2} = \dfrac{1}{1 + \dfrac{1}{1}}$, so the last digit 2 can be replaced by 1, 1

That is $\dfrac{25}{11} = [2,3,1,2] = [2,3,1,1,1]$

2) $0.57 = \dfrac{57}{100}$ $a_0 = 0$

$\dfrac{100}{57} = 1 + \dfrac{43}{57}$ $a_1 = 1$

$\dfrac{57}{43} = 1 + \dfrac{14}{43}$ $a_2 = 1$

$\dfrac{43}{14} = 3 + \dfrac{1}{14}$ $a_3 = 3, a_4 = 14$

Hence $0.57 = [0, 1, 1, 3, 14]$

Check

$$\cfrac{1}{1+\cfrac{1}{1+\cfrac{1}{3+\cfrac{1}{14}}}} = \cfrac{1}{1+\cfrac{1}{1+\cfrac{14}{43}}} = \cfrac{1}{1+\cfrac{43}{57}} = \cfrac{1}{\cfrac{100}{57}} = \cfrac{57}{100}$$

6.5   Metallic Means

Consider the equation from the previous section

$x^2 - nx - 1 = 0$

The solution is

$$x = \frac{n \pm \sqrt{n^2 + 4}}{2}$$

Take the positive root only,

$$\left(x = \tfrac{1}{2}n + \sqrt{n^2 + 4} \right)$$

We have seen that when n = 1, x = ϕ, the golden mean. In fact, n = 1, 2, 3...... form a series known as the *Metallic Means* (*ratios or constants*) labelled as golden, silver (n = 2), bronze (n = 3) .... copper, nickel ... That is, each metallic mean is a root of the quadratic equation

$x^2 - nx - 1 = 0$

And hence the continued fraction

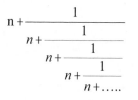

$$n + \cfrac{1}{n + \cfrac{1}{n + \cfrac{1}{n + \cfrac{1}{n + ...}}}}$$

The following table list out the metallic means for

n = 0 to 9.

n	0	1	2	3	4
Metallic Mean	1	$\dfrac{1+\sqrt{5}}{2}$	$1+\sqrt{2}$	$\dfrac{3+\sqrt{13}}{2}$	$2+\sqrt{5}$
Numeric Value	1	1.618	2.414	3.303	4.236

5	6	7	8	9
$\dfrac{5+\sqrt{29}}{2}$	$3+\sqrt{10}$	$\dfrac{7+\sqrt{53}}{2}$	$4+\sqrt{17}$	$\dfrac{9+\sqrt{85}}{2}$
5.193	6.162	7.14	8.123	9.11

Table 6.1 <u>Metallic Means</u>

Likewise for the equation

$$x^2 + nx - 1 = 0$$

the solution is

$$x = \frac{-n \pm \sqrt{n^2 + 4}}{2}$$

Neglect the negative root, $x = \frac{1}{2}\left(\sqrt{n^2 + 4} - n\right)$

For   n = 1, $x = \frac{1}{2}(\sqrt{5} - 1) = \phi_2$

A series can also be formed similar to the metallic means, let us name it as *Minor Metallic Means*. Other than a root of the above quadratic equation, each mean (ratio) is a continued fraction

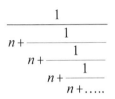

Table 6.2 list out the minor metallic means for n = 0 to 9. Note that the minor metallic mean x_2 is the fractional part of the corresponding metallic mean x_1. This can be observed from their continued fractions. More formally,

$$x_1 - x_2 = \tfrac{1}{2}\left(n + \sqrt{n^2 + 4}\right) - \tfrac{1}{2}\left(\sqrt{n^2 + 4} - n\right)$$

$$= \tfrac{1}{2}\left[n - (-n)\right]$$

$$= n$$

n	0	1	2	3	4
Minor Metallic Mean	1	$\dfrac{\sqrt{5} - 1}{2}$	$\sqrt{2} - 1$	$\dfrac{\sqrt{13} - 1}{2}$	$\sqrt{5} - 2$
Numeric Value	1	0.618	0.414	0.303	0.236

5	6	7	8	9
$\dfrac{\sqrt{29} - 5}{2}$	$\sqrt{10} - 3$	$\dfrac{\sqrt{53} - 7}{2}$	$\sqrt{17} - 4$	$\dfrac{\sqrt{85} - 9}{2}$
0.193	0.162	0.14	0.123	0.11

Table 6.2 <u>Minor Metallic Means</u>

Chapter 7

Fractals, Chaos, and Tessellation

7.1   Fractals

Suppose a series of changes in ratio or scale (uniform expansion or shrinkage) happens in a numerical sequence or geometric shape of an object. If the original pattern is retained after each transformation, then such entity or object is named as *fractal*. The term was coined by the Polish-origin mathematician Benoit B. Mandelbrot in 1975. (Later he published a book "The Fractal Geometry of Nature".) He realised the importance of *self-similarity* - many natural shapes display infinite sequences of motifs repeating themselves within motifs, on reducing or increasing scales. The former can be exemplified by the branching of tree and the latter the shell of a nautilus or snail. The pattern of the continued fraction of the golden section can be regarded as both fractal and self-similar.

It has been discussed that the golden triangle, golden rectangle, pentagon, pentagram and other geometric shapes related to φ show fractal properties. How about other shapes?

7.1.1   Cantor Set

In 1883, German mathematician George Cantor introduced the Cantor Set, also known as the Cantor Ternary Set. Start with a line segment, the first step is to remove the middle one-third of the line. Then remove the middle one-third of the remaining segments. The process goes on to infinity.

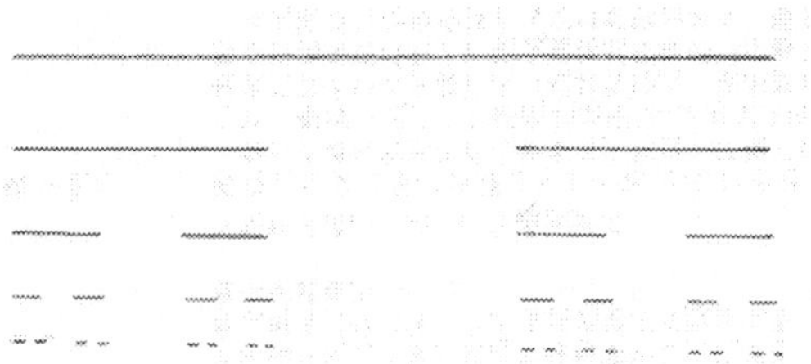

Figure 7.1 <u>Cantor Set</u>

Referring to the above figure, it can be seen that the total length removed is the sum of

$$S_n = \frac{1}{3} + 2\left(\frac{1}{9}\right) + 4\left(\frac{1}{27}\right) + 8\left(\frac{1}{81}\right) + 16\left(\frac{1}{243}\right) + \dots$$

$$= \frac{1}{3} + \frac{2}{9} + \frac{4}{27} + \frac{8}{81} + \frac{16}{243} + \dots$$

The series is a geometric one and each term is 2/3 of the previous one, i.e. the common factor, $r = \frac{2}{3}$, and $r < 1$.

For a geometric series with $r < 1$,

$$S_n = \frac{a}{1-r} \qquad \text{where } a \text{ is the first term}$$

Hence $\quad S_n = \dfrac{\dfrac{1}{3}}{1 - \dfrac{2}{3}} = 1$

That is, the whole length is removed and length left = 0.

But points are left behind. There are two types:

1) End points - countable (with divided segments)

2) Any points not in one of the middle thirds are not removed - uncountable.

So the cantor set contains infinite number of points.

The cantor set is a prototype of a fractal. It is self-similar, equals to 2 (reduced) copies of itself, contracted by a factor of 3.

Mandelbrot defined a term *fractal dimension* (also known as coastline dimension) D, which can be calculated by the formula

$\quad$ D = log a / log r

Where $\;$ a = number of pieces or copies
$\qquad\qquad$ (of the initial motif after one step)

$\qquad$ r = reduction factor

An object is said to be fractal if D is not an integer.

For the cantor set, a = 2, r = 3

Hence $\;$ D = log 2 / log 3 = 0.631

As D is not an integer, the cantor set is fractal.

7.1.2 Sierpinski Gasket

Another Polish mathematician Waclaw Sierpinski in 1919 devised another set of points using what we now call fractal subdivision. Start with an equilateral triangle (the *initiator*). Subdivide it into 3 smaller equilateral triangles with one on top the other two, leaving a void in the middle. (This is similar to removing an inverted reduced equilateral triangle in the middle.) Repeat with each of the reduced triangles. The result is the Sierpinski Gasket or Triangle. It is the set of points that remains inside the original triangle after repeating the above process an infinite number of times.

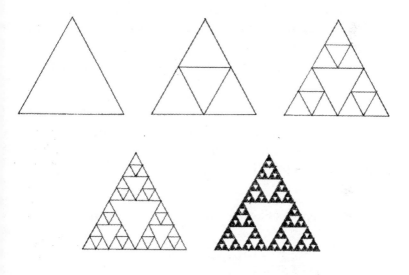

Figure 7.2 Construction of the Sierpinski Gasket

Figure 7.2 shows the first few steps of the construction. Note that the blacken areas in the last one represents clusters of the ever contracting triangles.

The number of copies, a = 3

Reduction factor, r = 2 (side of the reduced triangle is half the former one)

Hence D = log 3 / log 2 = 1.585

7.1.3   Koch Curve

Swedish mathematician Helge von Koch created a curve to demonstrate a mathematical oddity, which is then known as the Koch or von Koch curve. Start with a line segment. Divide it into 3 equal parts and construction an equilateral triangle at the middle part. Remove the base of the triangle. (This is similar to replace the middle part by two sides of an equilateral triangle of 1/3 the former length, forming a vertex.) Repeat the step with the reduced (now 4) line segments, and proceed on to infinity.

The result is a curve that contains no smooth sections. This was what von Koch wanted to present, a continuous curve that is not smooth and hence cannot construct tangent lines. Tangents are perquisite for the theorems of differentiation and calculus. The Koch curve came as an oddity.

The curve also looks like the contour of a coastline. It has

a = 4, r = 3

Hence D = log 4 / log 3
           = 1.262

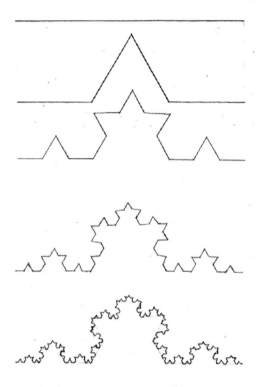

Figure 7.3 <u>Construction of the Koch Curve</u>

7.1.4 <u>Hilbert Curve</u>

The Hilbert curve is a continuous fractal space-filling curve, first brought forward by the German mathematician David Hilbert in 1891. It came as an alternative to the Peano curve which was the space-filling curve first described. Its main difference with the Peano curve is the number of copies in each step is 4 instead of 9 for the latter one.

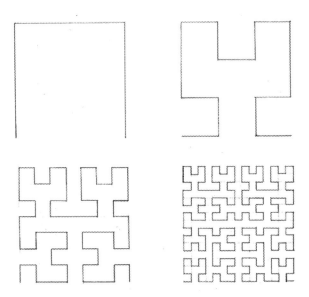

Figure 7.4 <u>Construction of the Hilbert Curve</u>

The Hilbert curve can be constructed in the steps below:

1) Start with a square without the bottom line, like an inverted U. The number of subdivided line segments (with respect to the original) in the nth step is governed by $s = 2^n - 1$.

2) $n = 2$, $s = 3$. Divide every line into 3 equal segments. Press down ("stamp") the top middle segment at a depth equals to the reduced length, and connect it to the remaining segments. Likewise, press down the bottom segments of the 2 sides. 4 inverted U are formed.

3) $n = 3$, $s = 7$. Replace each inverted U by the shape of step (2) with reduced size. The middle line segment at each of the 3 sides (at $n = 2$) remains there but at reduced scale.

4) Repeat step (3) with $s = 2^n - 1$, n approaches infinity.

The initiator of the Hilbert curve can be regarded as the shape at $n = 2$, rather than $n = 1$.

Note that $a = 4$, $r = 2^n - 1 / 2^{n-1} - 1$, which is different for different n. Hence D is not constant throughout.

For $n = 2$, $r = 3$

$$D = \log4 / \log 3 = 1.262$$

For $n = 3$, $r = 7 / 3$

$$D = \log 4 / \log(7/3) = 1.636$$

According to quotes from Wikipedia, the Hilbert curve gives a mapping between 1D and 2D space with preservation of locality. If (x, y) is a point within the unit square, and d the distance another point travels along the curve to reach that point, then points that have nearby d values will have nearby (x, y) co-ordinates. Nowadays Hilbert curve is widely applied in computer science. For example the range of IP (Internet Protocol) address can be mapped into a picture using the curve.

7.2   Chaos

Chaos can be regarded as the study of order in disorder, or vice versa. The questions are:

(1) Under what conditions will order (say periodic) becomes chaotic,

(2) Is there order (pattern) within disorder,

(3) Any relations with the golden ratio?

Chaos teaches us what can be predicted or forecast and what cannot. (Barnsley, 1988)

7.2.1   Butterfly Effect

Quantum mechanics postulates that at the sub-atomic scale, position and time cannot be determined exactly. It is only in terms of probabilities. In the macroscopic world, most objects do obey deterministic laws. However, some are still unpredictable like weather and throwing of dices. The reason is that we cannot measure the initial state of a system exactly.

If a system is sensitive to initial conditions, it exhibits the *Butterfly Effect* - "When a butterfly in Tokyo flaps its wings, the result may be a hurricane in Florida a month later". The system is said to be chaotic. Chaotic behaviour also obeys deterministic laws, but it is so irregular that it appears random.

In 1892, the French mathematician Henri Poincare introduced the concept of a *phase space*. This is a geometric representation of all possible motions of a dynamic system. Take a system of 2 variables at certain initial values. Their relative positions in movement are represented as co-ordinates on a 2 dimensional graph. The co-ordinates generate a curve as time proceeds. If the curve closes up into a loop, the 2 variables follow a periodic cycle. If the curve ends at a point, the system settles to a steady state. Different initial conditions may lead to the same final curve, the shape of the curve is known as an *attractor*. Any initial points lying outside the curve will be drawn into it (Figure 7.5). For system that settles to steady state, the attractor is a point. A closed loop attractor means the system is periodic, hence it is called an *oscillator*. Chaotic systems have their attractors in fractal shapes. They are called *strange attractors* (Stewart, 1995). What a chaotic system can be predicted is the shape of the attractor, which is not altered by the butterfly effect (initial conditions). American meteorologist E N Lorenz applied the theory to calculate weather patterns since 1963.

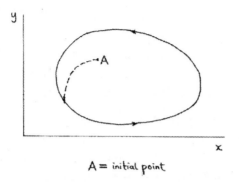

A = initial point

Figure 7.5 <u>A Closed Loop Attractor</u>

7.2.2   <u>Dynamics of Chaos</u>

Leo Kadinoff (1985) describes the fundamentals of dynamic systems as a sequence of points z_0, z_1, z_2 ….., each generate from the last one by the application of a function R;

i.e. $z_{j+1} = R(z_j)$

where z_j is a description of the state of the system at time t_j

One set of universal behaviour concerns maps in which z is real and R a function that obeys a kind of periodicity, such as

$$R(z) = z - \frac{h}{2\pi sin2\pi z}$$

where h is a parametric constant.

In this model, the oscillation of points around a circle, are represented by movements between the interval [0,1]. The paths produced by these maps are characterized by a *winding number*, defined as

$$w(z_0) = \lim_{j\to\infty} \frac{z_j}{j}$$

129

which describes the average number of revolutions travelled per step.

When $w = p / q$, a rational number, the system is stable. The orbit repeats itself after q iterations and cycling p times between [0,1]. Irrational winding numbers correspond to paths that never repeat.

For irrational orbits, their stability depends on h. If $h < 1$, the paths fill up the entire interval as $j \rightarrow \infty$. They are more or less stable. When $h = 1$, all orbits with irrational winding numbers have been destabilized, and further increase in h results in chaos. Orbits close to neighbouring rational orbits destabilized first. The golden ratio is the most irrational number and has the most distant neighbouring rational orbits, and is therefore the last to be destabilized.

In the chaotic regime, orbits may no longer fill up the entire circle, but bundle into narrow disconnected regions. These intervals are directly related to the Fibonacci numbers.

7.2.3   Predator - Prey Dynamics

Some predatory animals' population and survival depend on the availability of preys in their living zones. Take wolf as an example of the predator, there are many preys that it can feed on. For simplicity, we choose only one kind of prey, the deer. This forms a 2 variable predator-prey dynamic system.

The population of wolf depends on the existing population and the number of deer that it can feed on. So the population of wolf will not increase over a long period. If there is insufficient food, some will starve and die. Deer will not extinct totally because some can manage to escape the hunting. Under favourable climate, abundant supply of grass and the decreased number of wolf, their population will grow. Then the wolf will have more food available and the cycle repeats.

Hence there will not be a steady state where the 2 variable graph settles to a point, corresponds to the populations of both spies remain constant over time. The most likely outcome is an oscillation within a close loop like that in figure 7.5, where

x = population of deer, and

y = population of wolf.

If the periodic cycle breaks and say all the deer were killed by wolf or died from natural disaster, then the wolf will also extinct due to no food to prey on, or they have to migrate to another hunting ground, causing confrontation. That is chaotic.

7.2.4   Logistic Equation

The logistic model is different from the predator - prey system in that there is only one variable under consideration, the population in a place. It assumes that there is a maximum sustainable population. The variable x is the ratio of the present population to the maximum amount ($0 \leq x \leq 1$). The equation is given by

$$f(x) = kx\,(1 - x)$$

where $f(x)$ is an estimate of the population in the next time period, k is the growth rate constant.

The first part of the equation is $f(x) = kx$, the estimate population depends on the present population, more inhabitants will bear further more. However, when the population becomes too large, competition for the limited resources will cut down the growth by, kx^2, hence - kx^2. Here $f(x)$ is non-linear. Non-linear systems have greater tendency to become chaotic.

k may be regarded as a measure of the successive reproduction rate. If k is too low, $f(x) \rightarrow 0$ and the spies extinct. For intermediate k, the population settles to a steady state corresponds to $f(x_1) = f(x_2)\ldots.= f(x_n)$

If k increases, the population may oscillate between the larger and lesser values (Figure 7.6(a)). For larger k, the period splits into two, this process is known as *bifurcation* (Figure 7.6(b)). Further increase in k will lead to chaotic behaviour (Figure 7.6(c)).

It has been found that the ratio x remains positive until k reaches a critical value

$$k_c = \frac{3}{2}\sqrt{3} = 2.598$$

Above the critical value, the system is in chaos

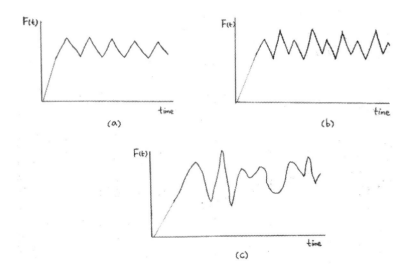

Figure 7.6 <u>From Periodic to Chaotic</u>

7.3   Tessellation

Tessellation of a flat surface is the filling up or tiling of it using one or more geometric shapes, known as tiles, with no overlaps or gaps. In Latin, tessellar means small square. It could mean a small, thin cubic piece of ceramic, stone or glass used to make mosaics. In 2 dimensional space, there are generally 2 types of tessellation: periodic, and aperiodic or non-periodic.

7.3.1   Periodic Tiling

A periodic tiling has repeated pattern(s). One or more modular shapes are formed and the surface is filled up by repeating the patterns. Usually, the building blocks are square, rectangles, triangles (45° or 60°), rhombus and regular hexagon. They are common in the mid-twentieth century in laying floor tiles. Figure 7.7 shows tiling by pure hexagon.

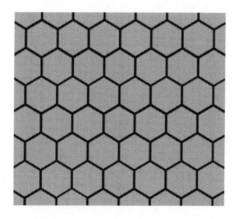

Figure 7.7 <u>Hexagon Tiling</u>
(Source: Pixabay)

Regular pentagon can also be used in periodic tiling, but it has to be combined with other polygons to eliminate the gaps.

7.3.2   Penrose Tiling

Penrose tiling is an aperiodic tessellation with pentagon, pentagram and decagon as the principal constituents. Therefore it is related to the golden geometry, and is self-similar and possesses 5-fold rotational symmetry (same outlook if rotated 1/5 of a circle). The tiling was first proposed by the English mathematician and physicist Roger Penrose in a 1974 paper. This was the first version, later labelled as P1. It composes of 6 prototiles (prototype shape). They are 3 pentagons with different colours, a pentagram, a "boat" (upper part of a star), and a diamond or thin rhombus.

The second version P2 has only 2 prototiles, a Kite and a Dart. The former is built-up from 2 golden triangles and the latter from 2 golden gnomons (Figure 7.8).

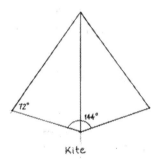

Kite

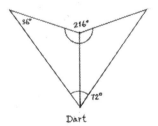

Dart

Figure 7.8 <u>Penrose Tiling (P2) : Kite and Dart</u>

5 kites form a decagon (Figure 7.9) and 5 darts form a star or pentagram (Figure 7.10). The decagon can be expanded to a star by adding 5 darts on the peripherals. Likewise, the star can be blown-up to a decagon by 10 kites.

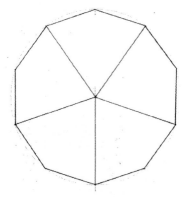

Figure 7.9 <u>A Decagon formed by 5 Kites</u>

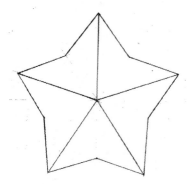

Figure 7.10 <u>A Pentagram formed by 5 Darts</u>

The third version (P3) is even more simple. It is based on 2 rhombi. One is with angles of 72° and 108°, called the Thick Rhombus.

Another has 36° and 144°, called the Thin Rhombus, as shown in figure 7.11.

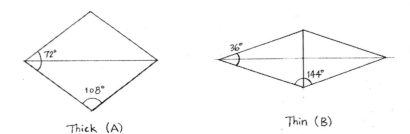

Thick (A)   Thin (B)

Figure 7.11 <u>Penrose Tiling (P3): Thick and Thin Rhombus</u>

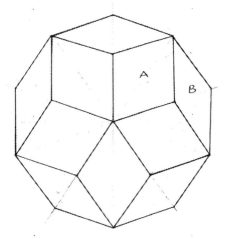

Figure 7.12 <u>Decagon formed from P3 Rhombi (Type 1)</u>

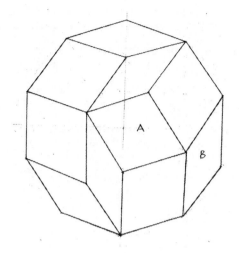

Figure 7.13 <u>Decagon formed from P3 Rhombi</u>
<u>(Type 2)</u>

The rhombi can be grouped to form overlapping decagons. Figures 7.12 and 7.13 depict the 2 types of grouping. Note that to lay a surface completely with these 2 decagons, the ratio of the former one to the latter is 1 : ϕ.

7.3.3   Other Tilings

The common star shape (figure 7.14) can be constructed from 5 rhombi with angles of 36°, 72°, 126°, 126°. The rhombus is formed by a golden triangle and a 72°, 54°, 54° isosceles triangle. This can be used as a stand-alone symbol, or combined with other polygons to form periodic or aperiodic tiling.

Tessellation can be extended to non-Euclidean space. One type is Hyperbolic Tiling, demonstrated extensively by the famous Dutch graphic artist Maurits C. Escher. In 1939, he made hexagonal tiling with reptiles as woodcut. The first hyperbolic tessellation was Circle Limit I (Figure 7.15), produced in 1958. It is an infinite regular repetition of the patterns in the hyperbolic plane, diminishing rapidly towards the edge of the circle. This is a representation of infinity on a 2 dimensional surface. All the "fish" are supposed to be of the same size and shape. They are distorted and appear progressively smaller towards the boundary. Other works of Escher can be found in the official website: www.mcescher.com.

The container in figure 7.16 is an example of tilings we may come across in everyday life. The tiling is on a 2 dimensional surface rolled up in a hollow cylinder.

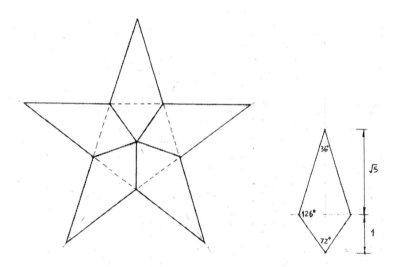

Figure 7.14 <u>A Star formed from the rhombus on the right</u>

Figure 7.15 <u>Hyperbolic Tiling: Circle Limit I</u>
<u>by M C Escher</u>
(Source: Wikipedia - Escher Circle Limit -
"free to modify, share or use commercially")

Figure 7.16 <u>Tiling on a Cylindrical Container</u>

Chapter 8

Conclusion

8.1   A Few More Explanations

We have gone through some "mysteries' about the golden ratio and try to explain them mathematically. Here are a few more that have left out or not fit into the former chapters.

8.1.1

The first is why the golden ratio ϕ is the most irrational number? Let us look at the numerical value of ϕ.

$$\phi = 1.6180339887499\ldots..$$

It can be seen that the pattern of decimal digits do not repeat themselves even at the 13th place. So this is one reason that ϕ is considered to be most irrational.

8.1.2

From $\phi^2 = \phi + 1$

$$\phi = \sqrt{1 + \phi}$$

Replace ϕ on the right hand side by the whole expression, and repeat to infinity,

$$\phi = \sqrt{1 + \sqrt{1 + \sqrt{1 + \sqrt{1 + \sqrt{1 + \ldots}}}}}$$

A continuous square root of 1.

8.1.3

In the book *Fivefold Symmetry*, Istvan Hargittai collectively wrote that "Regular pentagons and golden triangles may be endlessly dissected into smaller and smaller pentagons and golden triangles". The following 2 figures explain what this means. The pentagon in figure 8.1 is decomposed into 6 smaller pentagons and 5 triangular gaps. This form a half dodecahedral net (the technical term is 'development'). We will prove that the triangles are golden ones.

Let the side of a smaller pentagon be 1 unit. That is

$$AK = FB = 1$$

Δ OAB and Δ PFB are similar.

Therefore $\dfrac{AB}{FB} = \dfrac{OB}{PB}$

And $OB = OG + GP + PB$

By similarity, OG = GP (PG)

From section 4.2.2 and figure 4.8, we understand that PG is the radius of the circumscribed circle of the pentagon, and PB the radius of the inscribe circle; and

$$PB = (\sqrt{5} - 1)\, PG$$

Or $\quad PG = PB\,[1\,/\,(\sqrt{5} - 1)] = PB\,(\sqrt{5}+1)/4$

Therefore $\quad OB = 2PG + PB$
$$= PB\,(\sqrt{5}+1)/2 + PB$$
$$= PB\,(\phi + 1)$$

And $\quad AB = (OB/PB)\,FB$
$$= (\phi + 1)\,FB$$
$$= 1 + \phi_2 + 1$$

Since $\quad FB = AK = 1,$ $\qquad$ hence $\qquad KF = \phi_2$

As $\qquad HK = HF = 1,$

therefore $\qquad \Delta\,HKF$ is a golden triangle.

Note also $\qquad AB\,/\,FB = \phi + 1 = \phi^2$

That is, the fractal ratio of the pentagon is ϕ^2.

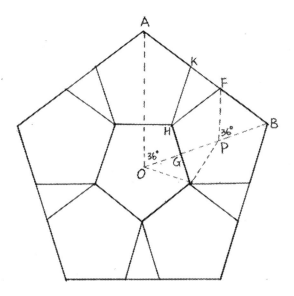

Figure 8.1 <u>Decomposition of a Pentagon</u>

For the golden triangle RST and golden gnomon LMN dissecting into pentagon and golden triangles, figure 8.2 is self-explanatory.

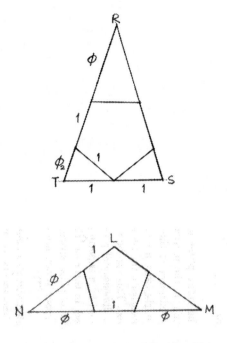

Figure 8.2 <u>Decomposition of Golden Triangle and Golden Gnomon</u>

8.1.4

Returning to section 6.4, it can be seen that the metallic mean in table 6.1 and the minor metallic mean in table 6.2 are reciprocals of each other. Another mystery? Let us look at the general solutions

Metallic mean $\qquad x_1 = \frac{1}{2}\left(\sqrt{n^2 + 4} + n\right)$

Minor metallic mean $\qquad x_2 = \frac{1}{2}\left(\sqrt{n^2+4}-n\right)$

$$1/x_1 = \frac{2}{\sqrt{n^2+4}+n}$$

$$= \frac{2\left(\sqrt{n^2+4}-n\right)}{\left(\sqrt{n^2+4}\right)^2-n^2}$$

$$= \frac{1}{2}\left(\sqrt{n^2+4}-n\right) = x_2$$

Hence x_2 is reciprocal of x_1.

8.2   Fascinating Mathematics

The author has shown that the golden ratio can easily been constructed from the 1 : 2 right angle triangle, with very simple tools like a piece of string / rope and a straight edge. The doubling of a unit can be achieved by the string also. In most ancient civilizations, people or tribes had been familiar with the concept of triangles and perpendicularity, that is right angle. Existing and ruined buildings and excavated foundations of their periods provide evidences. So are some deciphered texts and graphics. Therefore it is not so mysterious afterall that some ancient architecture composed of the golden ratio. It is just elementary geometry.

Before and during the Renaissance, the golden section had been regarded as the scared proportion after its promotion by the mathematicians and artists of that era. But whether it had been used by the famous artists in their paintings are still controversial. The applications in the later centuries are intentional and more direct.

The ratio of two consecutive Fibonacci numbers approaches to the golden section has been explained in section 6.2. The numbers lead to some mathematical puzzles. That goes to primordia, and fractal and self-similarity of spirals and golden geometry. The last topic discussed is the Penrose tessellation which is also interesting.

Therefore it is mathematics that is fascinating and not the golden section. Mathematicians create mathematical models to solve conceptual or abstract problems. They may just be metaphysics, logical or philosophical thinking with no immediate practical use. However, some models can be applied by scientists at a later stage to explain their experimental results or natural phenomena. They emerge as new scientific theories or discoveries and finally end up as new technologies.

8.3   Related Applications and Further Work

This section refers to contents in this book only. The concepts of chaos, fractal and self -similarities have been applied to contemporary organic architecture. Some building facades and outlooks have (seemingly) randomness beside familiar ordered patterns. Facades of modern skyscrapers exhibit fractal generation at different stages. Also, the use of Penrose tiling enriches our daily life.

Another immediate related application is to make use of the right angle triangle to produce various ratios. Given any right angle triangle with sides a : b, the figure below shows that the ratio of

$$\sqrt{a^2 + b^2} + a : b : \sqrt{a^2 + b^2} - a$$

can be obtained graphically. Some interesting ratios may be deduced here.

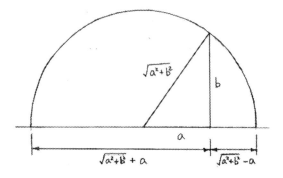

In particular when a = b, the ratio is

$$\sqrt{2} + 1 : 1 : \sqrt{2} - 1$$

Return to the basic problem of cutting a line in 2 segments. In section 5.1.4, we have seen that if the length of the whole line is taken as x + 1, x being the greater segment, the ratio of the greater to the smaller being twice the whole to the greater, then x = $\sqrt{3}$ + 1. In general, for the greater to the smaller segment being n times the whole to the greater with the above setting, the solution of x is given by the equation

$$x^2 - nx - n = 0$$

and $$x = \frac{1}{2}\left(\sqrt{n^2 + 4n} + n\right)$$

Hence we can form another set of means with different values of n.

The metallic and minor metallic means arise from the equation

$$x^2 \pm nx - 1 = 0$$

Is there any significance in the equation

$$x^2 \pm nx + 1 = 0 \qquad ?$$

Reference

'The Golden Ratio - the story of phi, the world's most astonishing number', Mario Livio, Broadway Books, 2002.

'Pi : a biography of the world's most mysterious number', Alfred S Posamentier, Prometheus Books, 2004.

'The Story of Architecture : From Antiquity to the Present', Jan Gympel, Konemann, 1996.

'Greek Architecture', R A Tomlinson, Bristol Classical Press, 1989.

'The Cambridge Introduction to Art - The Renaissance', Rosa Maria Letts, Cambridge University Press, 1981.

'The Maya - life, myth and art', Timothy Laughton, Duncan Baird Publishers, London, 1998.

'Art of the Andes from Chavin to Ica', 3rd edition, Rebecca R Stone, Thames & Hudson, 2012.

'Sacred Geometry - Philosophy and Practice', Robert Lawlor, Thames and Hudson, 1995.

'The Divine Proportion : A study in mathematical beauty', H E Huntley, Dover, 1970.

'The Golden Mean : Mathematics and the Fine Arts', C F Linn, Doubleday, 1974.

'The Golden Relationship : Art, Math, Nature', 2nd edition, M Boles, Pythagorean Press, 1987.

'Leonardo', Frank Zollner, Taschen, 2010.

'The Thirteen Books of Euclid's Elements, translated from the text of Heiberg', Thomas L Heath (introduction and commentary), Dover Publications, 1956.

'The King of Infinite Space : Euclid and his Elements', David Berlinski, Basic Books, 2013.

'Euclid : the creation of mathematics', Benno Artmann, New York Springer, 1999.

'A Mathematical History of the Golden Number', Roger Herz-Fischler, Dover Publication, 1998.

'Shapes - Nature's Patterns', Philip Ball, Oxford University Press, 2009.

'Nature's Numbers - discovering order and pattern in the universe', Ian Stewart, Weidenfeld & Nicolson, 1995.

'Mathematics - Powerful Patterns in Nature and Society', Harry Henderson, Chelsea House Publishers, 2007.

'Math Wonders to Inspire Teachers and Students', Alfred S Posamentier, Alexandria, 2003.

'Fivefold Symmetry', Istvan Hargittai (editor), World Scientific, 1992.

'Fearful Symmetry - is God a Geometer ?', Ian Stewart & Martin Golubitsky, Penguin, 1993.

'The fabulous Fibonacci numbers', Alfred S Posamentier & Ingmar Lehmann, Prometheus Books, 2007.

'Continued Fractions', C D Olds, The Mathematical Association of America, New Math Library, 1975.

'The Curves of Life', T A Cook, Dover, 1967.

'The Fractal Geometry of Nature', Benoit B Mandelbrot, Springer, 1983.

'Supercritical Behaviour of an Ordered Trajectory', L P Kadinoff, James Franck and Enrico Fermi, Institute Research Paper, 1985.

'Fractal Geometry in Architecture and Design', Carl Bovill, Boston: Birkhauser, 1996.

'Fractal Architecture: organic design philosophy in theory and practice', James Harris, University of New Mexico Press, 2012.

'The Emperor's New Mind: concerning computers, minds and the laws of physics', Roger Penrose, Oxford University Press, 1989.

'The Magic Mirror of M C Escher', Bruno Ernst, Taschen, 2007.